D1702425

SCIENCE, KNOWLEDGE, AND MIND

Science, Knowledge, and Mind

A Study in the Philosophy of C. S. Peirce

C. F. DELANEY

University of Notre Dame Press
Notre Dame London

Library of Congress Cataloging-in-Publication Data

Delaney, C. F. (Cornelius F.), 1938–
Science, knowledge and mind : a study in the philosophy of C.S. Peirce / C.F. Delaney
p. cm.
Includes bibliographical references and index.
ISBN 0-268-01748-4
1. Peirce, Charles S. (Charles Sanders), 1839–1914. 2. Science—Philosophy—History. 3. Knowledge, Theory of—History. 4. Philosophy of mind—History. I. Title.
B945.P44D39 1993
191—dc20 92-53743
CIP

To

Neil,

that he find 'the philosophical life'
personally fulfilling

Contents

Preface

Charles S. Peirce is the most important figure in the history of American philosophy to date for several interrelated reasons. First, with regard to that intangible mix of abilities that philosophers are wont to identify as constituting a great philosophical mind—that combination of analytic acumen and imaginative vision—Peirce is clearly America's candidate for a place on the roster of the great philosophers. Given his expertise both in mathematics and the physical sciences, he possessed to an extraordinary degree those technical skills philosophers so prize, and he mobilized these in the service of a systematic vision of the workings of the human mind and its place in the cosmos. Secondly, the interaction of his personal character traits with the social expectations of the nineteenth century made of his life the kind of personal tragedy that proves continually fascinating. The association of genuine creative genius with personal failure is the stuff, if not always of legends, at least of enduring curiosity. Finally, the issues in the philosophy of science, epistemology, and the philosophy of mind that were central to his philosophical project just happen to be ones that are still on center stage on the contemporary philosophical scene. His reflections still speak directly to our questions and, given the ineradicable egocentrism of human enterprises, this assures him of our continued attention.

For these and other reasons Peirce deserves a voice in our present conversation and the conversation would be enriched by his contribution. This book is an attempt to facilitate that entry. My interest in Peirce is much more than merely historical inasmuch as I feel that his version of pragmatism can function as an antidote to the anti-rational and anti-scientific strains that go under that label today. It seems to me that a pragmatism in the Peircean tradition represents a way of transcending many of the limitations of twentieth-century philosophy without turning our backs on its genuine advances over past ways of philosophizing.

This book has been many years in the writing and represents an attempt to piece together into a coherent vision themes I have explored over the years in numerous papers and addresses on Peirce. Sincere thanks are in order to those many individuals, both colleagues and students, who have patiently endured my numerous 'recalls to Peirce'. Special thanks are in order to Phil Quinn, who performed his customary and invaluable critical reading of the whole manuscript, and to Ernan McMullin and Richard Foley, who encouraged the project all along. I am also grateful for assistance of a more material sort to the Notre Dame Institute for Scholarship in the Liberal Arts for its generous summer support.

References to the Primary Texts

Throughout the body of the text Peirce's own work, both published and unpublished, will be referred to in a number of different sources. They will be indicated in the text parenthetically by the following abbreviations:

Collected Papers of Charles Sanders Peirce, ed. C. Hartshorne, P. Weiss, and A. Burks (Cambridge: Harvard University Press, 1931–1958). Since most of the references will be to this original collection, I will use the standard convention of volume number followed by paragraph number (e.g., 5.213), with no initial identifying letters.

Writings of Charles S. Peirce: A Chronological Edition, ed. M. Fisch et al., first four volumes completed (Bloomington: Indiana University Press, 1982–). References will be abbreviated as *W* followed by volume number and page number.

The Charles S. Peirce Papers microfilm collection (Harvard University Library, 1966). References will follow the numbering system developed by R. S. Robin in *Annotated Catalogue of the Papers of Charles S. Peirce* (Amherst: University of Massachusetts Press, 1967) and will be abbreviated MS followed by the manuscript number and the page number.

The New Elements of Mathematics, ed. C. Eisele (The Hague: Mouton, 1985). References are abbreviated *NE* followed by volume number and page number.

Charles S. Peirce: Contributions to "The Nation," 3 vols., ed. K. L. Ketner and J. E. Cook (Lubbock: Texas Tech University Press, 1975–79). References are abbreviated as *CN* followed by volume number and page number.

Charles S. Peirce: Letters to Lady Welby, ed. Irwin Lieb (New Haven: Whitlocks, Inc., 1953). References are abbreviated *LW* followed by page number.

I. *Introduction: Life and Work*

Charles S. Peirce was born in Cambridge, Massachusetts in 1839 into an intensely intellectual family. His grandfather was the librarian at Harvard and the author of a celebrated history of that university. His father, Benjamin Peirce, held the Perkins Chair of Mathematics and Astronomy at Harvard and was generally regarded as the foremost mathematician, if not man of science, in the country. Benjamin was the president of the American Association for the Advancement of Science and one of the founders of the National Academy of Sciences. In addition he helped found the Harvard Observatory and the Smithsonian Institution, and for a good deal of his life was associated with the United States Coast and Geodetic Survey, for which he served as superintendent for seven years. Charles' older brother, James Wells Peirce, also taught mathematics at Harvard and eventually succeeded to his father's professorship. An uncle, Charles Henry Davis, also lived in Cambridge and was the superintendent of both the *American Ephemeris and Nautical Almanac* and the Naval Observatory. This was no ordinary family.

Moreover, life in the Peirce household involved more than the family. Charles' home was one of the foci of intellectual life in a very intellectual city. The internationally renowned biologist and geologist, Louis Agassiz, was one of the closest family friends and lived just down the street. He, together with Charles' father and uncle, were the leading members of the Cambridge Scientific Club that met regularly in the Peirce home.The meetings of the Cambridge Astronomical Society were also regular events, as were the gatherings of the

Mathematical Club over which Benjamin Peirce presided. It was not without reason that Charles Peirce would later recall that "all the leading men of science, particularly astronomers and physicists, resorted to our house; so that I was brought up in an atmosphere of science" (*LW*, 37). Although science was obviously focal, the atmosphere in the Peirce household was much richer and more variegated than that provided by science alone. Longfellow, Lowell, Emerson, and Oliver Wendell Holmes were frequent visitors, as were some of the leading artists and actors of the time. It was in this intellectual and cultural hothouse that Charles S. Peirce grew up.

Even in this elite company young Charles did not grow up as one among many. Even against this comparison class his father regarded him as a prodigy, and, although there were three brothers and a sister, in a very special way Benjamin took personal charge of young Charles' education. In addition to the naturally to be expected tutelage in serious mathematics, he drew on Charles' early interest in complicated games, puzzles, and codes to develop in him extraordinary capacities of concentration and mental discipline. The methods were severe (including all night sessions with negative reinforcement) but the results were remarkable. Charles would later remark that he could work for sixteen hours without a break. Another uncle, Charles Henry Peirce, was the assistant to Eben Horsfod, the professor of chemistry at the newly established Lawrence Scientific School at Harvard, and he helped the young Charles set up a private laboratory in his home and guided him through the techniques of quantitative analysis. At age eleven Charles wrote a history of chemistry. At twelve he found his older brother's copy of Whately's *Elements of Logic* around the house, became fascinated, and mastered it in a week. Thereafter he began reading logic and philosophy under the guidance of his father. Although the young Charles did attend conventional schools in both Cambridge and Boston (with predictable and projectable tensions), I think it is safe to say that he received his formative education at home and that it was a formation long on independence of mind and mental discipline but short on social and personal responsibility.

This independence of mind was aptly encapsulated in his attitude upon entering Harvard in 1855, an attitude expressed in his intention "not to let Harvard interfere with his education." While at Harvard he studied seriously Schiller's *Aesthetic Letters* and worked on Kant's *Critique of Pure Reason* two hours a day for three years until he knew the whole book in both editions by heart. In a more public sense he managed to do some work in logic and retained his interest in chemistry. That he remained true to his intention upon entering was exhibited in the fact that in 1859 he graduated from Harvard near the bottom of his class with a less than positive attitude toward the college. He was, however, still regarded as the young prodigy of the time, and for him a scientific career was his and everyone else's expectation.

The year following his graduation from Harvard was profoundly educational but in the informal mode; he did some scientific field work in one area and submitted himself to some private tutelage in another. Both were to be influential. Another friend of his father, Alexander Bache, was superintendent of the Coast Survey (later the Coast and Geodetic Survey), and during 1859–60 Charles was a member of his party that did surveying in Maine and Louisiana, completing triangulations along the Atlantic and Gulf coasts. This was Peirce's initial contact with the institution which was at the time the chief scientific agency of the federal government and which was to be at the center of his scientific life for the next thirty years. Also during this year Charles apprenticed himself to another family friend, Louis Agassiz, to learn the techniques of zoological classification whose generalized employment would serve him well throughout his career. Measurement and classification were to be among his life-long preoccupations.

In spring of 1861 he entered the Lawrence Scientific School at Harvard to study chemistry. Chemistry was construed very broadly at the time, and in America it offered the best introduction to experimental science in general. Peirce distinguished himself in chemistry and chemical engineering, but his interests were developing more in the direction of logic and methodology. He graduated summa cum laude in 1863.

During his first year in "graduate school" Peirce regularized his association with the Coast Survey by accepting an appointment as a regular aide in July of 1861. Specifically, he was assigned as an aide to his father, who had been director of longitude determinations for the survey since 1852. Charles' job was as a computer on the project of the reduction of observations made over many years of occultations of the Pleiades by the moon, so as to determine the relative longitudes of American and European stations. These computations of theoretical astronomy not only gave Peirce experience with all the then known methods of computation but also provided him with the occasion to reflect more generally on the phenomena of observation, calculation, and probability. Moreover, it was work he was able to do in Cambridge while continuing to study and lecture in science proper, and the history and philosophy thereof. During this period he gave some very influential courses of graduate lectures at Harvard and at the Lowell Institute on the logic of science. The institution of graduate education did not yet exist, but these graduate lectures were thought of as "the germ of the graduate school." Peirce's career as a scientist reflecting on the foundations and methodology of science was well under way. It was during this period (1862) that he married Harriet Melusina Fay of a prominent New England family.

His success was publicly recognized by his election in 1867 to the American Academy of Arts and Sciences, to which academy he presented a series of five papers in logic. In the same year he was made an assistant in the Coast Survey, an appointment he was to hold until his resignation in 1891. He was assigned by his father, now superintendent of the survey, to the Harvard Observatory (which worked in conjunction with the Survey) to make astronomical observations under the direction of Professor Winlock. Two years later he was officially appointed as assistant in the Observatory. In his first years with the Observatory, he engaged in spectroscopic research involving auroral light and the observation of solar eclipses. His participation in the latter involved international travel for the purpose of observation. He continued to reflect on the methodology of his scientific work, specifically on the method of least squares used in astronomical reductions. This resulted in a paper "On the Theory of Errors of Observation" in 1870.

In 1872 Peirce became the principal investigator in two very important projects: first, the astronomical photo-metric researches conducted by the Harvard Observatory; and secondly, the gravity researches conducted by the Coast Survey. The Observatory had acquired a new astrophotometer and Peirce set out to determine the brightness or magnitudes of a whole range of stars. He embedded his own original observations in a discussion of other observations both historical and contemporary and justified his reduction of the various observations to a determinate figure. From his researches he drew some very general conclusions about the form and density of the Milky Way. These results were published in his "Photometric Researches" of 1878, for which he received professional acclaim.

At the same time he was in charge of the research conducted by the Coast Survey to determine the exact values of gravity at different stations throughout the world and from the variations determine the sphericity of the earth's shape. The determinations were made by pendulum swinging. The frequency of oscillation of a pendulum is a function of the intensity of gravity, and the variation of the latter with location provided a new way to determine the figure of the earth from surveys of gravity on the surface. In the process of making these determinations at various points around the globe, Peirce discovered a hitherto undetected source of error in such measurements brought about by the flexure of the tripod that occurred during oscillations of the pendulum. This discovery called into question all earlier measurements. His claim was at first resisted by the geodetic community but later corroborated and applauded.

The determination of these values of gravity required precise measurements of length, and for this purpose Peirce obtained from Berlin a standardized line-meter. He recognized that metallic bars were subject to change through molecular rearrangement and proposed a "spectrum meter" which would use the wave length of light as the standard unit. He constructed such a spectrum meter using the wave length of sodium light, a result later used by Rowland and acknowledged by Michelson and Morley. This creative scientific period culminated in Peirce's election in 1877 to the National Academy of Sciences.

Throughout this period Peirce continued to work on mathematics and especially logic. His theory of conformal map projections and his work on the four color problem grew out of his Coast Survey researches, but his principal formal work was in the area of logic. In fact, when asked to send a sample of his scientific work for consideration vis-à-vis his membership in the National Academy of Sciences, he sent only papers in logic and asked to be judged on the basis of these alone. This renders intelligible the nature of his only regular academic appointment.

In 1879 Peirce was appointed part-time lecturer in Logic at the newly constituted graduate institution, Johns Hopkins University. This did not interfere with his appointment to or work with the Coast and Geodetic Survey. For the next five years Peirce was active both as a working scientist and as a lecturer in logic and mathematics. In 1880 he was elected to the London Mathematical Society and a year later to the American Association for the Advancement of Science.

The Johns Hopkins years were extremely productive. His closest associate was probably J. J. Sylvester, the founder of the American Journal of Mathematics. Peirce taught courses not only in logic but in mathematics, and did serious work on the logic of relatives, the relative forms of quaternions, and the problem of continuity. It was also during this period that he developed independently of Frege what later came to be called quantification theory, a development that ensured his place in the history of logic. In 1883 he edited and contributed to *Studies in Logic*, published by Johns Hopkins. This volume contained papers by Peirce's students and associates, and most regard it as the first serious work in logic published in the United States. This year was also important personally, for during it he formally divorced his first wife (who had left him seven years earlier) and married Juliette Froissy, a somewhat mysterious Frenchwoman. In 1884 he was summarily dismissed from his position at Johns Hopkins, never again to hold an academic position.

Although Peirce continued his appointment with the Coast and Geodetic Survey, his professional life began to deteriorate. This was most likely occasioned by the death in 1880 of his father, who had functioned as a sponsor and patron in almost all facets of his scientific career. Without this buffer he was left to bear the consequences of his own idiosyncrasies and abrasive personality. Although Peirce continued his pendulum work and was for a time in charge of the Office of Weights and Measures, his relationship with the various superintendents of the Coast and Geodetic Survey went from bad to worse until his eventual resignation at the end of 1891.

In 1888 he had purchased a house (which he baptized "Arisbe") in a remote area of Milford, Pennsylvania to which he retired to think and write. He had a modest inheritance, but to make ends meet he translated papers for the Smithsonian Institution, wrote definitions for the *Century Dictionary*, wrote articles for the *Dictionary of Philosophy and Psychology*, gave occasional lectures, and for a time functioned as a consulting chemical engineer. He also worked on several textbook projects in mathematics that did not meet with success. His resources and health continued to deteriorate and, except for three or four lecture series back in Cambridge, he lived out the rest of his years in seclusion developing his ideas in logic and philosophy and attempting to bring his various insights together into some overall systematic statement. He died impoverished in 1914, leaving behind a voluminous collection of manuscripts, which were promptly purchased from his wife by his alma mater, Harvard University.

This picture of Charles S. Peirce as one of America's most distinguished working scientists of the nineteenth century is not meant to challenge the picture of him as a philosopher but to specify and complement it. During the whole period sketched above he was continuously working on what we might call purely philosophical issues. He, even given his penchant for fine-grained classificatory schemes, would question our sharp distinction. Peirce is a thinker—not unlike Aristotle, Galileo, Descartes, and Leibniz—for whom the relatively recent distinction between science and philosophy cannot be sharply drawn. Although he describes himself as growing up in the grips of science, from his earliest student days he was deeply immersed in philosophical projects both classical and contemporary and saw them as continuous with, rather than distinct from, his life as an experimental scientist.

Beginning with his undergraduate days, he had a consuming interest in Kant's philosophical project, i.e., the exploration of conditions of possibility of knowledge and of representation in general. His specific concern was with the theory of categories and the grounding thereof in formal logic. His conviction was that philosophy needed a new set of categories if it was going to comprehend reality as revealed by science. The conditions of possibility of knowledge had as its paradigm case the conditions of possibility of science. On the occasion of his election to the American Academy of Arts and Sciences in 1867 he presented to that body the fruits of his recent labor, three papers on logic. The third of these, "On a New List of Categories," Peirce later called his one real contribution to philosophy. A year later he published his justly famous "cognition series" in the *Journal of Speculative Philosophy*, in which he articulated his general theory of representation. These papers were the culmination of a very intense philosophical project that grew out of his reflections on Kant. They involved a fundamental critique of Cartesianism in all its guises, a critique informed by his developing views on the logic of science.

During the late 1860s Peirce became a close friend of the James family, and Henry Sr. and William were to have a profound influence on him. Together with William, Peirce founded a philosophy discussion group in Cambridge, which he called "The Metaphysical Club" (he was later to found a second such group in his Johns Hopkins years). Under this rubric Peirce, James, Chauncy Wright, Nicholas St. John Green, Oliver Wendell Holmes, John Fiske, and others met regularly to discuss fundamental questions in philosophy. If we are to lend credence to later reports, it was at these meetings that the fundamental ideas of pragmatism were born. Peirce worked continuously throughout this period on ideas that fermented in these discussions, and the product of his labors saw the light of day in his most famous series of papers, a series of six articles that appeared in *Popular Science Monthly* in 1878 under the general title "Illustrations of the Logic of Science." This series includes "The Fixation of Belief" and "How to Make our Ideas Clear."

If these early philosophical reflections could be said to concern themselves with the foundations of science, his later philosophical speculations could be described as extensions of science. In the 1880s, particularly after his retirement to Milford, he began to concern himself more with the broader metaphysical implications of and extrapolations from his account of science, i.e., the general picture of the universe which science, if not entailed, at least suggested. These speculative views were encapsulated in a series of papers that appeared in *The Monist* from 1891 to 1893. In 1902 he proposed a grand synthesis of all dimensions of his thought to the Carnegie Foundation, but they could not see their way to supporting the project. Nevertheless, interrupted by the occasional lecturing and dictionary writing he did for subsistence, Peirce continued to work on his grand philosophical project until his death in 1914.

It is the contention of this study that Peirce is most comprehensively viewed as a scientist-philosopher, where this hyphenated expression denotes an integrated attempt to understand the world and man's place in it using the best scientific information available. His more specific conception of philosophy was as a mode of inquiry *grounded in* and *reflective upon* mathematics and the experimental sciences. Some of his more comprehensive self-descriptions focus on the "reflective upon," whereas others focus on the "grounded in." From the former perspective he characterizes himself as fundamentally a logician, where logic is construed broadly as the exploration of the methodology of science (MS, 50.107), whereas from the latter perspective he describes his philosophical efforts as "the attempt of a physicist to make such conjectures as to the constitution of the universe as the methods of science may permit with the aid of what has been done by previous philosophers" (1.7). Moreover, he saw these two perspectives as being two moments, what one might call the analytic and the synthetic moments, in one overall philosophical project. From the foundations of science to speculative proto-science, it was a philosophical project with science at the center. As such, it was a project ideally suited to Peirce's rare combination of abilities.

A few preliminary remarks are in order with regard to the focus, the structure, and the tone of this particular study. The study will focus on Peirce's philosophy of science, theory of knowledge, and philosophy of mind in their interrelations. One might think that given the extremely broad range of Peirce's philosophical reflections this focus would track a rather narrow set of concerns, but such is not really the case. I would contend that this set of issues is at the absolute center of Peirce's overall philosophical vision and is the key to understanding his philosophy. These interrelated sets of issues are not just some among many; they are the issues that drive his philosophical engine.

A word or two must also be said about the development of Peirce's thought. Much of substance has been written about the overall development of his thought and about specific changes of view along the way. I acknowledge this and will in some instances allude to this developmental dimension when it bears on my specific issues. For the most part, however, I intend to present and explore his mature view as a coherent account of matters that persist throughout the changes. With regard to the issues I am concerned with, I think this approach is possible.

A final word by way of rationale for the order of exposition. I will start with his account of knowledge "writ large," i.e., his philosophy of science, and then move to the exploration of his general theory of knowledge and philosophy of mind as consonant with his view of science. While acknowledging that all these dimensions of his thought were being developed simultaneously, the chosen order seems to me to represent most perspicuously the motive force behind his thought. It is my contention that for Peirce science is at the center of his philosophical project, which project is best construed as the investigation of the logic of science, the conditions of possibility of science, and the speculative extension of science. Accordingly, it seems reasonable to get clear initially about his specific view of science narrowly construed, and then explore his more general views of knowledge and mind against this background. In any event, the structure of this study is an icon of its thesis.

In the conclusion I will make some specific suggestions about ways in which Peirce's perspective on this set of issues can reorient and illumine several contemporary philosophical projects. My tone not only in the conclusion but throughout may seem suspiciously contemporary, and I am very conscious of the fact that in the history of philosophy one must always be on one's guard against anachronism. But at any given time some historical figures really are our philosophical contemporaries. At the moment Frege and Sidgwick certainly have this status, and Peirce too, in the apt phrase of a recent commentator, "recognizably inhabits our philosophical world."[1] Our concerns really are his concerns, and little strain is involved in seeing him as addressing our questions, albeit with a slightly different and oftentimes more illuminating spin. In fact, a possibility worth reflecting on is that his natural resonance with our philosophical community may be part of the explanation of his dissonance with his own. Be that as it may, Peirce is certainly one nineteenth-century thinker with regard to whom one can both be historically responsible and yet find philosophically relevant.[2]

II. *Peirce's Account of Science*

One associates the phrases "the structure of science" and especially "the conditions of possibility of science" with Immanuel Kant, and Peirce's commitment to the Kantian project is certainly well-documented. He states that during his formative years as an undergraduate at Harvard "he devoted two hours a day to the study of Kant's *Critic of Pure Reason* for more than three years until I knew almost the whole book by heart and critically examined every section of it" (1.4). In the early 1860s he single-mindedly devoted himself to a rethinking of the Transcendental Analytic which culminated in his 1867 paper "New List of Categories," and he then described Kant as "now acknowledged everywhere as the master of philosophy" (*W*, 1.241). To invoke an earlier idiom, Kant was for Peirce *The Philosopher*, and Peirce's own philosophical reflections are so related to his mentor that, in his own words, "I am forced back again to Kant and find myself unable to take a single step until I have defined somewhat the principles upon which his philosophy is founded" (*W*, 1.241).

But as in most cases of genuine philosophical inspiration or tutelage, there is a substantial difference between the master and the pupil, a difference which is best understood along the lines of a critical rethinking of the Kantian project rather than a misunderstanding or critique of it. The project at issue is the two-fold one of exploring the logical structure of science and exhibiting the conditions of its possibility. It is not so much in their conception of this philosophical project that they fundamentally differ but rather in their respective conceptions of *science*, a difference which has its ultimate ground in the centrality of the notions of 'history' and 'community' in Peirce's overall philosophical orientation. These features of his general orientation disposed Peirce to what I will call a concrete as opposed to an abstract conception of science. In contrast to the focal conception of science being that of a static set of propositions (Euclidean geometry or Newtonian mechanics), for Peirce the focal conception of science is that of a socio-historical process of inquiry with a specific structure. This fundamental shift in perspective on science gives a quite different tone to the project of exploring its logical structure and exhibiting the conditions of its possibility.

1. The Nature and Preeminence of Science

a. The Nature of Science

For Peirce there are two quite different traditional conceptions of science: the first is the characterization of science primarily as a systematic or organized body of knowledge; and the second is the characterization of it primarily as a method of knowing. The former he viewed as a rather "shallow cut" which captured only "the fossilized remains of science" (MS, 614.7). The latter he saw as a "deeper cut" which was surely on the right track but which in its traditional articulations was compromised by an excessively individualistic and not sufficiently dynamic conception of methodology. He drew on his own experience as a scientist, his knowledge of the history of science, and his expertise as a methodologist of science in an effort to characterize the concrete reality that was living science in contrast to some abstract specification of some feature thereof.

This socio-historical picture of science receives its ultimate generalization in his construal of science as a "mode of life." Applauding Bacon's vision of science (while demurring at many particulars), Peirce proposes the following definition of the word "science":

> For him [Bacon] man is nature's interpreter; and in spite of the crudity of some anticipations, the idea of science

> is, in his mind, inseparably bound up with that of a life devoted to single minded inquiry. That is the way in which every scientific man thinks of science. That is the sense in which the word is to be understood in this chapter. Science is to mean for us a mode of life whose single animating purpose is to find out the real truth, which pursues this purpose by a well-considered method, founded on thorough acquaintance with such scientific results already ascertained by others as may be available, and which seeks cooperation in the hope that the truth may be found, if not by any of the actual inquirers, yet utimately by those who come after them and who shall make use of their results. It makes no difference how imperfect a man's knowledge may be, how mixed with error and prejudice; from the moment that he engages in an inquiry in the spirit described, that which occupies him is *science*, as the word will be here used. (7.54–55)

This concrete characterization he also extends to our conceptions of the particular branches of science. He views a particular science, e.g., chemistry, as "no mere word manufactured by some academic pedant but as a real object, being the very concrete life of a social group constituted by real facts of inter-relation" (7.52). From Peirce's perspective, then, when we speak of science in general or of some particular science what we are concretely talking about is a community of inquirers extended over time with a unity of purpose and method which enables the product to be much more than the sum of the products of the individual contributors. It is in this spirit that Peirce returns to the question of definition and states that "if we are to define science, not in the sense of stuffing it into an artificial pigeonhole where it may be found again by some insignificant mark, but in the sense of characterizing it *as a living historical entity*, we must conceive it as that about which men such as I have described busy themselves" (1.44).

From the fact that his primary conception of science as a mode of life is thus concrete and global, it doesn't follow at all that he thinks that a detailed account of the logical structure of scientific method is somehow inappropriate. On the contrary, the general features of Peirce's analysis of the logic of scientific method as involving abductive, deductive and inductive phases is one of the hallmarks of his philosophy. On this analysis, the abductive phase is that concerned with the original generation and recommendation of explanatory hypotheses; the deductive phase has to do with the logical elaboration of the hypothesis; and the inductive phase bears on the confirmation or falsification of the hypothesis by future experience. It is important to keep in mind, however, that these are construed as phases in a process of inquiry and are ultimately to be understood in terms of their role in that overall process.[1]

A more detailed look at this tripartite structure of the method of scientific inquiry will enable us to get a clearer picture of his understanding of "the logic of science." While as a first approximation we might think of *abduction* on the model of inference to the best explanation, Peirce's own account is more nuanced than this characterization suggests. What counts as "best" calls for disambiguation. In his more fine-grained account of the abductive phase of scientific inquiry Peirce distinguishes two moments. The first moment is simply the origination of those conjectures that will make up the list of possible explanations of the phenomena under consideration. This 'discovery' moment is a matter of the creative imagination of some people. Some people upon being confronted with a perplexing array of phenomena are able to imagine structures such that were such operating or structuring the behavior of what was operating, the phenomena in question would be rendered intelligible. Peirce thinks of this creative ability in terms of natural instinct and does not think that it can be reduced to strict formulae or rules of procedure.

But if we are to think of abduction as the proposal of a hypothesis for consideration, there is more to abduction than discovery narrowly construed. The second moment takes its rise from the fact that there may well emerge several suggested hypotheses that equally 'explain' the facts. Accordingly, if we are to get on with the task of scientific explanation we must select from our list of possible explanations those we are seriously to consider and, furthermore, effect a preference ordering of these. This moment of the abductive process is rule-governed, and Peirce does articulate a set of regulative principles that should guide this process of antecedent theory choice.

The first such regulative principle that Peirce articulates for the second moment of the abductive phase of inquiry is that we should "follow the rule that one of all admissible hypotheses which seems the simplest to the human mind ought to be taken up for examination first" (6.532). The operative phrase here is "seems simplest to the human mind," so it is most important to get clear about the notion of simplicity being invoked here. In commenting on Galileo's use of this notion, Peirce specifies what he himself means by "simple" much more firmly:

> That truly inspired prophet had said that, of two hypotheses, the *simpler* is to be preferred, but I was formerly one of those who, in our dull self-conceit fancying ourselves more sly than he, twisted his maxim to mean the *logically* simpler, the one that adds least to what has been observed, in spite of three obvious objections: first, that so there was no support for any hypothesis; secondly, that by the same token we ought to content ourselves with simply formulating the special observations actually made; and thirdly, that every advance of science that further opens the truth to our view discloses a world of unexpected complications. It was not until long experience forced me to realize that subsequent discoveries were every time showing that I had been wrong, while those who had understood the maxim as Galileo had done early unlocked the secret, that the scales fell from my eyes and my mind awoke to the broad flaming daylight that it is the simpler hypothesis in the sense of the

> more facile and *natural*, the one that instinct suggests, that must be preferred; for the reason that unless man have a natural bent in accordance with nature's, he has no chance of understanding nature at all.... I do not mean that logical simplicity is of no value at all, but only that its value is badly secondary to that of simplicity in the other sense. (6.477)

In the first instance, then, Peirce maintains that we should select for serious consideration those hypotheses that seem most natural to us, those with which we feel most comfortable, those that "naturally recommend themselves to the mind and make upon us the impression of simplicity—which here means facility of comprehension by the human mind—of aptness, of reasonableness, of good sense" (7.220). Just as he thought of the initial creative ability in terms of a natural instinct for guessing right, he invokes this same notion of natural instinct as operative in the moment of antecedent theory selection. We will take a closer look at this alleged instinctive ability when examining his views on the conditions of possibility of science.

While still speaking of this notion of simplicity as a criterion of antecedent theory selection, Peirce invokes another kind of consideration which spills over into other rules for this stage in scientific inquiry:

> This rule has another advantage which is that the simplest hypotheses are those of which the consequences are most readily deduced and compared with observation; so that, if they are wrong they can be eliminated at less expense than any others. (6.532)

The point here is that those hypotheses simplest for us are the easiest ones for us to work with, so starting with them will enable us to get on with the inquiry most efficiently. Inasmuch as they involve views with which we feel most comfortable, we can quickly see what each entails and imagine how to go about testing them. Should they prove promising, we may well be on the road to the true explanation; should they prove to be false starters, at least they can be eliminated with less trouble and so speed us on our way down other avenues. In this perspective on simplicity we can see encapsulated Peirce's *dynamic* (as opposed to static) and *social* (as opposed to individualistic) view of scientific inquiry. Simplicity so understood is not a logical feature of a theory at a time but a property a theory has in virtue of certain roles it can play at given historical stages of the inquiry process. Moreover, the justification of the criterion is not merely in terms of the individual inquirer but in terms of the overall functioning of the community of inquirers.

These points are brought to the fore when Peirce moves from this discussion of simplicity to the much broader rules the discussion suggests:

> This remark at once suggests another rule, namely, that if there be any hypothesis which we happen to be well-provided with means for testing, or which, for any reason, promises not to detain us long unless it be true, that hypothesis ought to be taken up early for examination. (6.533)

Here Peirce is situating the criterion of simplicity in the broader context of what he called "the economies of research," an area of investigation he carved out in his 1876 paper entitled "A Note on the Theory of the Economy of Research."[2] Here, under the general rubric of *economy*, Peirce organizes several regulative principles that bear on this second moment of the abductive process. His meta-comment on the set of principles he is to formulate puts the matter succinctly: "What really is in all cases the leading consideration in Abduction is the question of Economy—Economy of money, time, thought and energy" (5.600). The term "economy" is used very broadly here to range over the various scarce human resources that are invested in our cognitive endeavors, and the moral is that our principal concern ought to be to realize the best possible cognitive return on our investment. It is from this general consideration of economy that a set of rules for the selection of a hypothesis in this second moment of abduction is generated.

The first rule is a straightforward application of the general principle: "If any hypothesis can be put to the test of experiment with very little expense of any kind, that should be regarded as a recommendation for giving it precedence" (7.220). In the same vein there is the recommendation that that hypothesis should be preferred which "can be the most readily refuted if it is false" (1.120). A second rule, which could be seen as a corollary of the first, recommends that one hypothesis should be preferred over another if one could not test the latter without doing almost all the work required to test the former but not vice versa (7.93). A third rule, expressed in the jargon of the game of billiards, maintains that hypothesis should be preferred which if false would give the best "leave," i.e., whose residuals would be the most instructive with reference to the next move to be explored (7.221). Again in the same vein Peirce invokes the game of Twenty Questions and recommends the preference of that hypothesis which would halve the number of possible explanations (7.220). A fourth rule enjoins us on the basis of economy to prefer (all else being equal) the broader of two hypotheses on the grounds that the illumination it will shed on the general inquiry will be greater whether it is true or false (7.221). Taken together these rules amount to the injunction that at the abductive stage of inquiry the investigator should do a cost-benefit analysis of the various paths along which she can proceed and allow the result of this analysis to determine her choice.

The reason for this injunction Peirce thinks is obvious. The number of suggested explanations may be considerable and the process of verification to which they must be submitted before they can count as items of even likely knowledge is very expensive. Our resources of time, energy, and money—and the ideas which might be had with that time, energy, and money—are very limited. Accordingly, "economy would override every other consideration even if there were any other serious considerations; in fact there are no others" (5.602).

A point to be emphasized is that in this account of the criteria for antecedent theory choice in terms of a general theory of the economics of research, Peirce clearly construes these criteria both *historically* and *socially*. One hypothesis is to be preferred over another not in terms of its intrinsic merits or its likelihood of being true but in terms of the role it can play in a process of inquiry which is aimed at truth in the long run. An hypothesis is recommended now to the degree that its pursuance at this point in time would move the inquiry along most efficiently. Peirce's invocation of the game of Twenty Questions is instructive. In playing this game, a line of questioning recommends itself not in terms of the likelihood it will "hit upon" the correct answer immediately but in terms of the role this line of questioning will play in hastening the eventual convergence on the answer. Secondly, the justification of these rules is not in terms of the individual investigator but in terms of the community of investigators to which she belongs. The hypothesis recommended to any individual at a time may not at all be the one most likely to enable her to attain the truth but rather the one the present investigation of which will most efficiently ensure the eventual attainment of truth. Given the state of the inquiry, an individual investigator may be rationally constrained to spend her days eliminating some unlikely possibilities. Accordingly, if we are to think of abduction in terms of inference to the best explanation, 'best' is to be understood not in terms of 'likely to be true' or even 'providing the most comprehensive explanations and/or predictions now' but rather in terms of 'efficiency in moving the inquiry along in its convergence on the truth as quickly as possible'. Of course, it may be best not to think of abduction in terms of inference to the best explanation but rather to save this current model for the explication of the emergence of a given preferred hypothesis at the end of the inquiry process.

The second phase of scientific inquiry, *deduction*, is much less distinctive and thus easier to grasp. It consists in the logical analysis of the hypothesis quite independent of the evidence for and against it in the attempt to generate all sorts of necessary and probable experiential consequences which would follow from it (7.203). It involves two steps: first the logical analysis of the hypothesis as given with the aim of rendering it as distinct, complete, and consistent as possible; and secondly, the derivation from it of certain 'predictions' concerning experienceables which could count for or against its being true (6.471).

Peirce distinguishes two kinds of deduction which not only adds specification to the account but also illumines the two different steps in this stage of the inquiry process:

> Deductions are of two kinds, which I call *corollarial* and *theorematic*. The corollarial are those reasonings by which all corollaries and the majority of what are called theorems are introduced. If you take the thesis of a corollary,—i.e., the proposition to be proved, and carefully analyze its meaning, by substituting for each term its definition, you will find that its truth follows in a straightforward manner from previous propositions similarly analyzed. But when it comes to proving a major theorem, you will very often find you will have need of a *lemma*, which is a demonstrable proposition about something outside the subject of inquiry; and even if the lemma does not have to be demonstrated, it is necessary to introduce the definition of something which the thesis of the theorem does not contemplate. (7.204)

The specific point here is to draw a distinction between the purely mechanical notion of deductive explication whereby the conclusion is "deduced directly from propositions established without the use of any other construction than one necessarily suggested in apprehending the enunciation of the proposition" (*NE*, 4.288) from the more creative process of proving some surprising conclusion from our premises by "imaginatively experimenting on the image of the premises" (*NE*, 4.38). Thinking specifically about the mathematical discovery, Peirce draws the distinction this way: "A Corollarial Deduction is one which represents the conditions of the conclusion in a diagram and finds from the observation of this diagram, as it is, the truth of the conclusion; a Theorematic Deduction is one which, having represented the conditions of the conclusion in a diagram, performs an ingenious experiment upon the diagram, and by the observation of the diagram so modified, ascertains the truth of the conclusion" (2.667). Theorematic deduction, then, involves real imagination, whether by way of experimentation on this diagram or by the situation of our deduction in a wider context of truths or procedures on which it draws. Both kinds of deduction are involved in this second phase of scientific inquiry in specifying the hypothesis under consideration, and the second surely will be central to deducing from the hypothesis experiential consequences the occurrence or non-occurrence of which will confirm or falsify the hypothesis.

This brings us to the third stage of inquiry, *induction* or confirmation. "Induction" as Peirce uses the term is not to be understood simply in terms of the relation of individual cases to a general law but in terms of the *role* such a logical relationship plays in the inquiry process in general:

> The only sound procedure for induction, whose business consists in testing an hypothesis already recommended by the retroductive [abductive] procedure, is to receive its suggestions from the hypothesis first, to take up the predictions of experience which it conditionally makes, and then try the experiments to see whether it turns out as it was virtually predicted in the hypothesis that it would. Throughout an investigation it is well to bear prominently in mind just what it is we are trying to accomplish in that particular stage of the work at which we have arrived. Now when we get to the inductive stage what we are about is finding out how much like the truth our hypothesis is, that is, what proportion of its anticipations will be verified. (2.755)

Hence, the principal role of the inductive relationship in the inquiry process is that of confirmation or falsification. That is, the role of the relationship between propositions describing individual instances to the relevant laws or theories is not that of the latter being derived from them, but rather of these instances, having been predicted by the theory, functioning either to confirm or to falsify it. It is through the inductive phase that the speculative flight of scientific imagination is continually monitored by experience.

Much more needs to be said about this inductive monitoring because it is in virtue of its role in the inquiry that Peirce sees himself able to construe science as a self-regulating and self-corrective process:

> Induction is the experimental testing of a theory. The justification of it is that, although the conclusion at any stage of the investigation may be more or less erroneous, yet a further application of the same method must correct the error. (5.145)

His point is that the method of testing which this inductive phase of inquiry involves is such that "if it be persisted in long enough it will assuredly correct any error concerning future experience into which it may temporarily lead us" (2.769).

This picture of confirmation or inductive monitoring clearly involves a considerable commitment to a kind of empiricism; not an empiricism about the origin of ideas but about their testing. The commitment necessary is expressed in Peirce's famous 'pragmatic maxim'. One can be confident that the deliverances of experience are at least the right kinds of things to be invoked in the confirmation or falsification of theories because the very meanings of theoretical terms are given in terms of conditionals referring to future experiences. In order for an hypothesis to be meaningful, i.e., to be admissible into the scientific inquiry process, it must have conceivable empirical consequences that distinguish it from other hypotheses and thus allow it to be confirmable or falsifiable by future experience.

The exact formulation of the pragmatic maxim is even more empirically restrictive than this account suggests:

> All reasonings turn upon the idea that if one exerts certain kinds of volition, one will undergo in return certain compulsory perceptions. Now this sort of consideration, namely, that certain lines of conduct will entail certain kinds of inevitable experiences is what is called a 'practical consideration'. Hence is justified the maxim belief in which constitutes pragmatism; namely, *In order to ascertain the meaning of an intellectual conception one should consider what practical consequences might conceivably result by necessity from the truth of that conception; and the sum of these consequences will constitute the entire meaning of the conception*. (5.9)

The strong thesis is the claim that if we are to grasp the meaning of any of our hypotheses, we need only examine conditionals of the form, "If a certain action were performed, certain experiential consequences would be observed." Nothing else is involved; the list of all conditionals of this kind exhausts the meaning of the hypothesis. This theory of meaning for theoretical terms unambiguously directs one to the *kinds* of things that must function in confirmation and falsification.

It is most important to note that when Peirce is here talking about future experiences he is not talking about private mental events but about intersubjectively available experimental phenomena: "It is not *my* experience but *our* experience that has to be thought of" (5.402, n.2). About this he is quite explicit: "Indeed, it is not in an experiment, but in *experimental phenomena*, that rational meaning is said to consist; when an experimentalist talks of a *phenomenon* such as 'Hall's phenomenon', 'Zemman's phenomenon' and its modification 'Michelson's phenomenon', or 'the chessboard phenomenon', he does not mean any particular event that did happen to somebody in the dead past, but what surely will happen to everybody in the living future who shall fulfill certain conditions" (5.425). These are the kinds of future experiences that count.

Peirce realizes that the notions of intersubjective experience, reproducible phenomena, and replications of observations are more complicated notions than might at first appear because strictly speaking "no two observers can make the same observations. . . [and even] the observations which I made yesterday are not the same which I make today" (*W*, 3.55). Hence when we talk about the agreement of future observations, we are talking about observation *types* rather than *tokens* and about the agreement of trained observers with regard to them. Even at this fundamental level of observation, science's commitment to generality and to social agreement manifests itself. These general points being understood, the issues that remain to be clarified with regard to confirmation have to do with numbers (how many instances?), distribution (what kind of sampling?), and the economic constraints on confirmation.

When we commit ourselves to the program of testing our hypotheses by the deliverances of experience, we are certainly aware of the fact that any given hypothesis will range over many more cases that can possibly come under scrutiny. His view is that this lack of fit will not discredit the procedure provided our inductions satisfy the conditions of 'predesignation' and 'random sampling'. First, it is perfectly obvious that if we simply look over the facts to find agreements with our theory, "it is a mere question of ingenuity and industry how many we shall find" (7.775). For this reason Peirce insists that the character of the prediction "be specified in advance of, or at least, quite independently of any examination of the facts" (2.789). Secondly, in addition to predesignating the character to be tested for, the inquirer must take steps to ensure that the instances he will specifically examine constitute a random sample of the class of instances in question. The investigator should "collect on scientific principles a fair sample of the *S*'s, taking due account in doing so of the intention of using its proportion of members that possess the predesignate character of being *P*; this sample will contain none of those *S*'s on which the retroduction was founded; the induction then presumes that the values of the proportion among the *S*'s of the sample of those that are *P*, probably approximates within a certain limit of approximation, to the value of the real probability in question" (2.758). Peirce's contention is that so long as we are careful to predesignate the cases that will count and follow the procedures of fair sampling we can be assured that the continued application of inductive procedure will reliably eliminate false theories and thus by indirection recommend the true one. The self-corrective features of the method ensure its convergence on the truth; thus we will eventually discover whether or not reality has those features our theory ascribes to it. The ultimate justification for this picture of scientific inquiry, particularly the 'assurance' promised, involves Peirce's exploration of the conditions of possibility of science, a matter taken up later in this chapter.

Economic considerations also come into play in the inductive phase of inquiry: "Throughout the process of verification the exigencies of the economy of research should be carefully studied from the point of view of its abstract theory" (7.90). As our knowledge of a given domain becomes more and more complete, the cost of additional confirmation and precision goes up considerably, such that "it does not pay ... to push the investigation beyond a certain point of fullness and precision" (1.122). A balance must be struck between the desire for completeness and the cost of attaining it. The finite nature of human cognitive capacities even in the long run implies that for every line of inquiry there is "an appropriate standard of certitude and exactitude, such that it is useless to require more and unsatisfactory to have less" (1.85). Peirce does not view this mundane observation as compromising his realism but just as circumscribing it.

It is important to attend to the fact that on Peirce's account of scientific method the three inference structures—abduction, deduction, and induction—are to be construed dynamically in their interrelations as phases in the overall inquiry process. Hypotheses are suggested, elaborated, and eliminated in such a way that the total inquiry process is both creative and self-corrective through its various phases.

b. The Preeminence of Science

Scientific method, however, is not the only model of cognitive inquiry. Peirce characterizes three other general types (which he calls 'the method of tenacity', 'the method of authority', and 'the a priori method') that have had historical periods of dominance and which continue to vie with science for our allegiance. 'Tenacity' is the simple and direct strategy of taking our given belief and "constantly reiterating it to ourselves, dwelling on all which may conduce to that belief, and learning to turn with contempt and hatred from anything that might disturb it" (5.377). 'Authority' is the practice of eschewing self-determination of belief and deferring to some authority, usually a church or state, to define the belief that we should adopt and defend (5.37).[3] Both these methods make the determination of belief a matter of 'will', whether of the individual or the organization. The 'a priori method' is more intellectual in nature, allowing those beliefs to be fixed which are 'agreeable to reason', intuitively obvious, self-evident, or simply those with regard to which we find in ourselves a natural inclination to believe (5.382).

What are we to make of this list of four methods and the arguments for the preeminence of science by way of the exclusion of the other three? The form of such an argument presupposes that the list is exhaustive; but there would seem to be other familiar methods, e.g., the hermeneutical method or the method of psychoanalysis, not represented on the list. By way of response to this kind of concern, it is important to appreciate the high level of generality on which Peirce is operating when he articulates these four methods of fixing belief. The practices for fixing belief he is exploring are those where the final determination is a matter of will, that where it is a matter of taste, and that where it is a matter of adjustment to an independent reality. Other more specific methods we might recognize are not alternatives on this level but rather specific strategies that could be followed in any of Peirce's four ways. Scientific method is not in contrast with, for example, the hermeneutical method but with various ways in which the hermeneutical method might be employed.

But even at this high level of generality there are these other methods, methods which in some circumstances can be very effective. In this context Peirce argues that what he calls scientific method is not just one model of cognitive inquiry on the same footing with these others. For him science is clearly privileged; it is *the* model of cognitive inquiry if we really are interested in solidifying belief, proceeding rationally, and ultimately attaining the truth. It is instructive to look at the kinds of reasons Peirce brings to bear on this issue of the preeminence of science so understood.

The principal argument he puts forward for the privileged status of scientific method focuses on the notion of solidifying belief and has a socio-historical character. He views cognition as a kind of adaption-oriented human activity, with 'science' being one of the competitive cognitive models. The 'best' cognitive model will be the one which enables its employer most effectively to adapt to his environment and hence survive. The considerations that Peirce brings to bear to show that the scientific model is the best are pragmatic in nature and are both social and developmental. The failure of the models of tenacity and authority are traceable to their insularity. The former model would be effective in fixing belief only if one were a hermit; Peirce's simple objection is that "the social impulse is against it" (5.378). Given our social nature, this method will break down in practice. The second model would be effective in a closed society, but gives way in the face of "wider social feeling" (5.381). The inevitable glimpses of other societies or other times, while they may have an initial hardening effect, sow the seeds of doubt which, once they take root, cannot be overcome by this model itself. The third model, the a priori or self-evidence model, although it has the air of intellectuality, really reduces to a matter of taste and as such will not be effective in enabling its preferred beliefs to survive and prosper in a public and diverse intersubjective domain. The scientific model alone, according to Peirce, seems to have the resources to fix belief in such a way as to give its practitioner any founded hope for effective long-run cooperative interaction with his expanding community and shrinking environment. In short, it is the only model which can fix belief effectively.

In addition to this empirical or pragmatic argument in defense of the preeminence of science, there are other argument strands in the same "Fixation of Belief" paper that can be distinguished from it. First, there is what might be called "the argument from the hypothesis of reality," which hypothesis Peirce spells out in some detail:

> There are Real things, whose characters are entirely independent of our opinions about them; those Reals affect our senses according to regular laws, and, though our sensations are as different as are our relations to the objects, yet, by taking advantage of the laws of perception, we can ascertain by reasoning how things really and truly are; and any man, if he have sufficient experience and he reason enough about it, will be led to the one true conclusion. (5.384)

The hypothesis states that the structure of reality does not depend on the will of any individual or group of individuals, yet can be uncovered by rational inquiry due to the experiential effect the real has on our senses and thus on our opinions. The point being made is that this hypothesis is in harmony with scientific method and thus "no doubts of the method necessarily arise from its practice, as is the case with all the others" (5.384). The last phrase signals an ambiguity in the argument. The point seemingly being made is that scientific method *alone* is consistent with the hypothesis and that this fact shows its preeminence over the other methods. This would be the case if this 'hypothesis of reality' were a presupposition of all methods of inquiry and hence of the other three as well. In fact, the next consideration he puts forward contains the seeds of such a generalized argument. Here, however, the hypothesis of reality is introduced simply as a presupposition of scientific method, and if it is only this, a lack of harmony between it and the other methods surely doesn't count against them. One demands a harmony only between a given method and *its* fundamental conception.

Secondly, there is what might be termed "the argument from error":

> This is the only one of the four methods which presents any distinction of a right and a wrong way. If I adopt the method of tenacity, and shut myself out from all influences, whatever I think necessary to doing this is necessary according to that method. So with the method of authority: the state may try to put down heresy by means which from a scientific point of view seem very ill-calculated to accomplish its purposes; but the only test *on that method* is what the state thinks; so that it cannot pursue the method wrongly. So with the a priori method. The very essence of it is to think as one is inclined to think! ... But with the scientific method the case is different. I may start with known and observed facts to proceed to the unknown; and yet the rules which I follow in doing so may not be such as investigation would approve. The test of whether I am truly following the method is not an immediate appeal to my feelings and purposes, but, on the contrary, itself involves the application of the method. Hence it is that bad reasoning as well as good reasoning is possible; and this fact is the foundation of the practical side of logic. (5.385)

One might take the point here to be the simple one that it is only with regard to scientific method that we have the resources to discover that we are using the method incorrectly. Accordingly, it is the only method that can amend its own structure or correct its own employment.[4] Upon closer examination, however, the point seems to be a deeper and more elusive one. What he says is that it is the only method that admits of a distinction between its right and wrong employment. Whatever is done according to the other three methods is by definition the right thing to do, whereas employing scientific method admits of misemployment. Peirce may well think that this shows not merely that science is the best method but that it is the only method, because for something truly to be a method at all it must admit of misemployment. But the text itself is only suggestive. What is clear, however, is that Peirce has a distinctive understanding of science as a mode of inquiry and reasons for thinking that it is the preeminent mode of inquiry.

2. The Conditions of Possibility of Science

It is this understanding of scientific inquiry that informs Peirce's fundamental project of exhibiting the conditions of possibility of science. There are two features of science so understood that seem to call for philosophical grounding. First, science is from Peirce's own perspective only one model of cognitive inquiry, and one which has not always been dominant and whose continuance is by no means inevitable. Secondly, it is a mode of inquiry whose objective validity, whose possibility of attaining the truth, has no obvious guarantee. Hence, what are the conditions of possibility of its continuance and flourishing, on the one hand, and of its objective validity on the other? As I will reconstruct Peirce's project, then, it will have two facets: first, the articulation of certain qualities of inquirers and institutions necessary to sustain the process; and secondly, the articulation or positing of certain features of our world necessary to guarantee its objective validity. Together these will constitute the conditions of possibility of science as we know it.

a. The Conditions of Possibility of the Development and Continuance of Science as a Mode of Inquiry

Since there are other models of cognitive inquiry that have had historical periods of dominance and which continue to vie with science for our allegiance, the development and continuance of science as we know it depends on the persistence of certain interrelated norms, practices, and institutions that characterize its members, define its structure, and delineate its boundaries.

Starting from the bottom up, Peirce locates the distinctiveness of science as practiced and grounds its continued development in certain *virtues* being embodied in the individual members of the community of investigators. Sometimes he speaks of these as 'norms' or even as a 'code of honor' (*MS*, 615.14) but most frequently simply as 'moral factors':

> The most vital factors in the method of modern science have not been the following of this or that logical prescription—although these have had their value too—but they have been the *moral factors*. (7.87)

And the particular moral factors he specifies are *the love of truth*, *the sense of community*, and *the sense of confidence*.

The first of what Peirce calls 'moral factors' seems initially to be completely uncontroversial: "The first of these has been the genuine love of truth and the conviction that nothing else could long endure" (7.87). This apparently straightforward claim, however, masks some complexities. Peirce is not claiming that inquiry qua scientific is either disinterested or presuppositionless. On the contrary, he is well aware of the many interests that can motivate a given line of scientific inquiry. His claim is that for the truly scientific mind the love of truth is the dominant one. Sometimes the contrast he has in mind is between this purely cognitive motivation and other more or less noble motives such as fame, money, or social welfare (8.143), but most often the contrast is between two cognitive attitudes, namely, the mind-set of the inquirer and the mind-set of the pedagogue. He sees the scientist as the one whose dominant driving interest is in the search for truth wherever it may lead, whereas he sees the pedagogue, whether teacher or preacher, as one whose dominant interest is in organizing and communicating what he already knows. He characterizes the difference concretely as the distinction between the laboratory mind and the seminary mind. This contrast between the quest for truth and the elaboration and dissemination of belief runs deep into the human character, and it is a contrast the Peirce sees 'writ large' in the difference between the spirit of modern science and that of the Middle Ages.

Nor does Peirce think that scientific inquiry is presuppositionless. Any given scientific inquiry is not only conducted against the background of the 'established scientific verities' of the moment but also against the background of more general metaphysical assumptions which guide our orientation to the world (7.82). It is our attitude toward these presuppositions that can be either scientific or not. If the quest for truth is dominant, these background presuppositions are never regarded as beyond question. Although, as entrenched, the presumption is in their favor, if the direction of the inquiry seems to call for their revision, then such a revision must be regarded as a real option in the interest of truth.

The second moral factor, the sense of community, is more complicated and attracts more of Peirce's attention. In addition to making the obvious points about the requirement of intersubjectivity of evidence imposed by the social character of scientific investigation, he goes on to explore the deeper commitments of self-sacrifice and self-abnegation involved in the enterprise which is science:

> The method of modern science is social in respect to the solidarity of its efforts. The scientific world is like a colony of insects in that the individual strives to produce that which he cannot himself hope to enjoy. One generation collects premises in order that a distant generation may discover what they mean. When a problem comes before the scientific world, a hundred men immediately set all their energies to work on it. One contributes this, another that. Another company, standing on the shoulders of the first, strikes a little higher until at last the parapet is attained. (7.87)

Mixing his metaphors between 'a colony of insects' and a 'company of troops', Peirce makes the point that the life of science is essentially that of an historical community that is teleological in structure. The development and continuance of this life depends on the social sense becoming supreme through the individual investigators developing those virtues that will enable them to subordinate their own satisfaction to the long range goals of the community.

For Peirce these personal virtues that constitute the sense of community are not merely an extrinsic support for the life of science but are essentially tied to the very logic of scientific method:

> It can be shown that no inference of any individual can be thoroughly logical without certain *determinations of his mind* which do not concern any one inference immediately. (5.354)

And these 'determinations of mind' involve the individual's viewing his particular inferences not just as part of the larger set of *his own* inferences but in terms of their role in that ongoing inquiry the proper logical subject of which is the historical community. It is with this in mind that Peirce articulates what he calls the three logical sentiments, namely, "interest in an indefinite community, recognition of the possibility of this interest being made supreme, and hope in the unlimited continuance of intellectual activity" (2.655) as indispensable requirements of logic. The life of science demands the transcendence of both selfishness and skepticism through the active hope that rational cooperative effort will in the end prevail.

This leads to the third moral factor undergirding the development of science, namely, the sense of confidence. He thinks that a sense of confidence is particularly crucial in an enterprise that proceeds by the method of conjecture and refutation. With regard to our specific proposed explanations we are clearly going to be wrong more often than we are right, so it is important that we continue to view our proximate failures in terms of their contribution to the long-range effort. He sees this confidence as characteristic of scientists: "Modern science has never faltered in its confidence that it would ultimately find out the truth concerning any question to which it could apply the check of experiment" (7.87). This attitude of mind implies both the correctness and eventual completeness of science, and takes the form of the action-guiding hope that the indefinite application of scientific methodology will lead to success in the long run.

Given Peirce's pragmatism, it should not be surprising that he saw these moral factors not as private internal states but as embodied in a coordinated set of *practices* constitutive of the scientific community. He saw this single-minded concern for truth together with the sense of community and confidence manifested behaviorally in scientists' "unreserved discussions with one another" and their "availing themselves of their neighbor's results," which practices developed into the constraints to make experiments replicable and evidence intersubjectively available. It is the network of these practices that is seen to constitute a given science as "a real object, being the concrete life of a social group constituted by real facts of interrelation" (7.52).

Moreover, the *institution* so constituted is not only the judge and repository of past results but even more importantly the locus of those criteria of evaluation of present programs and prognoses of future ones that afford dynamic continuity to the scientific enterprise and give it an identity over time that transcends any individual or group of practitioners. With regard to these matters Peirce's involvement was not merely abstract and theoretical but concrete and practical. He saw as one of the most fundamental problems for the scientific community the rational determination of "how, with a given expenditure of money, time and energy, to obtain the most valuable addition to our knowledge" (7.140). In response to this problem he worked out specific criteria in an area he called 'The Economy of Research' that would function in the rational assessment on a cost-benefit basis of proposed research programs so as to optimize the allocation of limited resources in its pursuit of long-range goals. It was his conviction that if these, or criteria like them, were to be adopted by the funding arm of the scientific community, judgments otherwise unprincipled would come under the purview of rational criteria designed with long range success in mind.

It is this network of specific norms, practices, and institutions that distinguishes science as a cognitive way of life from other modes of fixing belief and is the ground of its continual development. We are now in a position to see what precisely it is about a cognitive process so constituted that enables it to fix belief most effectively. On Peirce's account the other three cognitive models have their effectiveness undermined for the same reason; the ever widening sphere of social interaction inevitably introduces factors of diversity which erode the insular consensus that forms their respective bases. Having no cognitive resources to deal with dissonance, confidence wanes and effective action breaks down. The scientific model, on the other hand, makes a virtue out of this vice by incorporating the social factors from the beginning and by building in mechanisms to take account of the diversity of opinion and plurality of perspectives. The data base is allowed to be as broad as it can be and presumed to be diverse; the cognitive machinery is designed for continual adjustment to new inputs so that successive equilibrium points are found which are stable and can function as guides for action. The social factors are neutralized as threats by being incorporated as contributors. Our beliefs are guaranteed a dynamic stability. But this, of course, is only part of the story. What reason do we have to believe that a method effective for stabilizing belief bears any presumptive relation to the production of beliefs characterizable as *objective* and *true*?

b. The Conditions of Possibility of Objectivity and Truth and Science

What is it about the scientific way of proceeding that guarantees or at least renders probable its cognitive success? What must reality be like in order for science so construed to attain objectivity and truth? Since Peirce is looking at science as an historical process, his dynamic version of the Kantian project involves two important differences. First, the question is no longer "What must the world be like for these specific claims of science to be true?" but rather "What must the world be like in order for this mode of inquiry to ultimately attain the truth?" The notion of validity *here and now* is replaced by the notion of validity *in the long run*; however, it is still far from obvious why procedures effective for belief stabilization should have anything to do with success understood as objectivity or truth. Secondly, Peirce's characteristic fallibilism is introduced, severing the tie between 'necessarily true' and 'objectively valid'. The classical concern with demonstration and necessary truth is abated, with the result that objective validity is tied to the process of inquiry rather than to a priori structures. It is in this spirit that Peirce displaces the Kantian question "how are synthetic *a priori* judgements possible?" in favor of the more general one, "how are synthetic judgements *in general* possible?" (5.348). To come to grips with these questions we will have to take a closer look at his account of scientific method.

The deductive phase as 'objectivity' or 'truth' preserving seems relatively unproblematic, but both the abductive and inductive phases involve evidential gaps that call for some kind of bridging if our confidence in science's objectivity and truth is to be grounded. Given the number of possible explanations abstractly available for any set of phenomena, what account can be given of our ability to come up with antecedently plausible explanations of our world; and secondly, what justification can we give for our confidence that continued application of confirmation procedures will lead to truth in the long run? Both creative foci, that of discovery and that of confirmation, involve evidential leaps that call for some kind of justification.

Peirce's account of the factors that metaphysically ground abduction and induction falls somewhere between a likely story and a transcendental argument. He takes the history of science as the phenomenon to be explained, and he sees it as exhibiting a progressive development of empirical adequacy, predictive success, manipulative control, and explanatory power. It is to account for this multi-faceted progressive development that Peirce sees fit to postulate certain special features to the inquirer's relation to the world that would render the *rate* and *degrees* of the various kinds of cognitive success intelligible.

This strategy is most pronounced in his explication of the abductive or 'discovery' phase of scientific inquiry, the phase which Peirce clearly thought was the most important and which most called for explanation:

> What sort of *validity* can be attributed to the First Stage of inquiry? Observe that neither Deduction nor Induction contributes the smallest *positive item* to final conclusion of the inquiry. They render the indefinite, definite; Deduction explicates; Induction evaluates; that's all. Over the chasm that yawns between the ultimate goal of science and such ideas of man's environment as those coming over him during his primeval wanderings in the forest ... we are building a cantilever bridge of induction, held together by scientific struts and ties. Yet every plank of its advance is first laid by Retroduction [Abduction] alone ... and neither Deduction nor Induction contributes a single new concept to the structure. (6.475)

It's not that deduction and induction are not crucial to scientific inquiry; it's that all the concepts that figure in the propositional content of scientific explanations—all the 'positive items' as he calls them—must *enter* through the abductive process.[5]

The issue is the legitimacy of the inputs, the grounding of the process whereby the ideas which are to be the building blocks of explanations are initially introduced into scientific inquiry. The thought seems to be that if we can't have confidence in the material introduced, how can we have confidence in the security of the completed building? But this very thought should set off warning bells. The building metaphor certainly suggests epistemic foundations, and isn't one of the more salient features of Peirce's overall philosophical orientation his critique of foundationalism? Reflection on this point can lead us deeper into Peirce's conception of science.

Peirce's pragmatism is forward-looking, not backward-looking; the engine of justification is self-correction, not foundations. In contrast to Aristotle and Descartes, his point is *not* that if there is going to be any justification at the end, then there must be solid justification at the beginning, or that the process of inquiry is at best justification preserving and most frequently justification diminishing. On the contrary, Peirce's own view is that whatever errors we start with or are led into at the beginning will be corrected over time by the process of inquiry itself, so that the justification of the conclusion is genuinely built through the process.

Why, then, is he concerned with the grounding of the initial inputs—for his concern with the validity of abduction *is* an epistemic concern about the initial introduction of explanatory ideas. The answer, I believe, is to be found in his reflections on the history of science. Peirce thinks that we have made genuine *progress* in scientific understanding that calls for explanation at many different levels. Moreover, he thinks that the question "How was man ever led to entertain a correct theory?" is prior to the questions having to do with "How he came to believe it" or "How is his belief justified?" Speaking of scientific theories that we acknowledge to be true, he asks:

> How was it that man was ever led to *entertain* that true theory? You can't say that it happened by chance because the possible theories, if not strictly innumerable, at any rate exceed a trillion ... and therefore the chances are too overwhelmingly against the single true theory in the twenty or thirty thousand years during which man has been a thinking animal ever to have come into man's head. (5.591)

The point is that when he looks at the history of science and sees this history as exhibiting a multi-faceted cognitive progress, he feels such could not be explained on the basis of merely random inputs. Given the de facto time span presented us, without some special assumptions bearing on the antecedent reliability of the inputs, it would be extremely improbable that "even the greatest mind would have attained the amount of knowledge which is actually possessed by the lowest idiot" (2.753). Hence the point is *not* that scientific explanation would be impossible without firm foundations but rather that the de facto success *rate* of science—its progress in the time allotted—would be unintelligible given purely random inputs or initial ideas that had no grounding in the order of things or in logical principles. Hence his speculation about the 'validity of abduction'.

As we have seen, Peirce's concern with this issue of the introduction of ideas into the inquiry process is structured by his division of the abductive phase into two different moments each with a quite different kind of grounding or rationale. The first moment bears on discovery properly so-called, namely, the origination of those conjectures which will make up the list of possible explanations of the phenomena under consideration. Here Peirce ruminates about man's ability to come up with antecedently plausible explanation candidates from an almost infinite number of abstract possibilities. In the end he posits a 'natural instinct for guessing right' as the first part of any account of the de facto historical rate of success of science. This instinct, he says, is "strong enough not to be overwhelmingly more often wrong than right" (5.173) and, given the iteration of the process, this should be enough to get the right explanation on the list to be considered. The second moment has to do with the selection from our list of antecedently plausible hypotheses those we are to take as serious candidates for investigation and the order in which we are to take them. It is with regard to this moment that Peirce develops his rules of economy, but even at this level there is an irreducible role for our 'natural instinct for guessing right' in our preference ordering of those hypotheses that should inform our research programs (7.220).

What is the metaphysical ground of this cognitively crucial natural instinct? For Peirce it is not left as a bare posit but is accompanied by a likely story of its existence and functioning in creative scientific minds. His story is broadly evolutionary in nature. Given the obvious presence of survival instincts in the rest of the animal kingdom, Peirce's first thought is that it is not unreasonable to believe that we too have those instincts necessary for the effective continuance of our distinctive mode of life (6.476). He sees the ability to guess right as having obvious adaptive value and hence as a clear candidate for natural selection. In particular, he views a rudimentary grasp of certain fundamental principles of mechanics as crucially important to certain organic practices necessary for survival.

> The great utility and indispensability of the conceptions of time, space and force even to the lowest intelligence are such as to suggest that they are the results of natural selection. Without something like geometrical, kinetic and mechanical conceptions, no animal could seize his food or do anything which might be necessary for the preservation of the species ... As that animal would have an immense advantage in the struggle for life whose mechanical conceptions did not break down in a novel situation (such as development must bring about), there would be a constant selection in favor of more and more correct ideas of these matters. (6.418)

It is important to note that Peirce is here talking about the ability to guess *right* (the impetus toward "more and more correct ideas of these matters"); he thinks that it is only beliefs that are on the right track that will have staying power and developmental fecundity in our ever changing circumstances. Since it is correct beliefs that will have survival value, it should not be at all surprising that we—the survivors—should have this ability to guess correctly to a considerable degree.[6]

There is another chapter to the evolutionary story, one having to do with how we came to have this ability in the first place. If we are operating with a Cartesian picture of the inquirer's relation to nature such that the inquiring mind is, as it were, outside nature looking in trying to guess at the laws which describe its structure, then our apparent success rate would indeed be mysterious. But if we assume that the inquiring mind is constituted by nature's evolving development, its affinity with its object becomes less mysterious:

> If the universe conforms with any approach to accuracy to certain highly pervasive laws, and if man's mind has been developed under the influence of those laws, it is to be expected that he should have a *natural light* or *light of nature* or *instinctive insight* or genius tending to make him guess those laws aright or nearly aright. (5.604)

Quite ironically he finds the ultimate ground for these Cartesian notions (natural light or light of nature) in a decidedly anti-Cartesian picture of mind: "Our minds having been formed under the influence of phenomena governed by the laws of *mechanics*, certain conceptions entering into these laws become implanted in our minds so that we readily guess at what the laws are" (6.10). Being nature's products, we have ready access to her secrets.

Finally, this general story is particularized for the major players in the history of science: "Galileo appeals to *il lume naturale* at the most crucial stages of his reasoning; Kepler, Gilbert and Harvey—not to speak of Copernicus—substantially rely on an inward power, not sufficient to reach the truth by itself, but yet supplying an essential factor to the influences carrying their minds to the truth" (1.80). Evolutionary speculation is tied down to the specific historical phenomenon of the occasional creative genius who is indispensable to the progress of science.

Peirce now turns to the inductive phase of the inquiry process to explore the question of its metaphysical grounding. Is his confidence that continued application of confirmation procedures will lead to the truth in the long run merely an unfounded hope, or can it be shown to have some objective foundation? What must the world be like so as to render intelligible the success of this self-monitoring feature of scientific inquiry and thereby the long-run objectivity of science as a whole? Specifically this is a question about the alleged self-corrective dimension of science while more generally it is a question about the justification of induction. For, if the confirmation procedures of science are going to be viewed as playing this crucial role in this movement toward truth over time, then we must have reason to believe that the inductive sampling that functions in the confirmation stage is not destined to be misleading and can function as a reliable guide to the structure of the real.

That Peirce believes that the inductive procedure has this feature is quite clear. He views it as a procedure which "if steadily persisted in must lead to true knowledge in the long run of cases of its application whether to the existing world or to any imaginable world whatsoever" (7.207). Induction is for him self-monitoring in that its continued use will uncover the mistakes in its earlier uses such that by this process of purification truth will be eventually attained. He sees this as quite independent of any particular features of the world. In fact, he can't even imagine a world in which such a procedure would not be reliable:

> If men were not able to learn from induction it might be because, as a general rule, when they have made an induction the order of things would then undergo a revolution ... But this general rule would itself be capable of being discovered by induction; and so it must be a law of such a universe that when this was discovered it would cease to operate. But this second law would itself be capable of discovery. And so in such a universe there would be nothing which would not sooner or later be known; and it would have an order capable of discovery by a sufficiently long course of reasoning. (5.352)

If even in such a demonically contrived universe inductive procedures would be reliable, surely they would be reliable in any ordinary universe—ours in particular.

It is important to re-emphasize that he is not looking at the confirmation process (the inductive phase of inquiry) in terms of the degree of warrant any specific theory has in terms of specific test results, but in terms of the long-run effect of continued empirical testing. The self-correctiveness of science crucially involves the monitoring role of the inductive phase, but the inductive phase does not supply the successive hypotheses; it functions only to eliminate. Better hypotheses are generated by the whole process of inquiry over time; it is a matter of genuinely promising suggestions continually subjecting themselves to elaboration, prediction, and possible elimination. The individual scientist is viewed as a member of an historical community whose bond of unity is the employment of this self-corrective method. The logical subject of the inquiry is the *scientific community over time*. The continual monitoring at the inductive phase occasions conceptual revision through the abductive phase, with the overall process of inquiry resulting over time in a more and more adequate picture of the world.

Peirce's acceptance of this picture of science is tied to his construal of the monitoring phase (induction–confirmation) specifically in terms of statistical inductions. Specific predictions as to the character of our world are derived from the hypothesis under investigation, and then our world is checked for this character. 'Samples' are drawn from our world to see if it has the characteristics we suppose it to have. Obviously there is some initial evidence for this character, but if controlled sampling belies its presence, continued sampling should reveal which of the other proposed characteristics really map onto our world. Why this should be the case brings us to the heart of Peirce's pragmatic realism:

> An endless series must have some character; and it would be absurd to say that experience has a character which is never manifested. But there is no other way in which the character of that series can manifest itself than while the endless series is still incomplete. Therefore, if the character manifested by the series up to a certain point is not that character which the entire series possess, still, as the series goes on, it must certainly tend

> however irregularly toward becoming so; and all the rest of the reasoner's life will be a continuation of this inferential process. This inference does not depend on any assumption that the series will be endless, that the future will be like the past, or that nature is uniform, or any other material assumption whatsoever. (2.784)

His idea is that inductive inference is basically an inference from part to whole, and its validity depends simply on the fact that parts do make up and constitute the whole. In confirmation we are basically involved in drawing samples from a population, and if the frequency with which some relevant property is distributed over the individuals of that sample does not correspond to its frequency of distribution over the population, the discrepancy is sure to become apparent as the sampling process is extended over the long run. To resist this line of thought is to entertain a conception of the population or the whole which will never manifest itself in the samples or the parts. But to entertain this is to conceive of truth as possibly transcendent, reality as possibly incognizable, both of which Peirce thinks he has good reason to reject.

3. 'Truth' and 'Realism' in Science

Given the socio-historical conception of scientific inquiry Peirce has articulated, one might think that the natural capstone of such an account would be a straightforward consensus theory of truth and a social-construction view of reality, with all the contingency and relativity such views involve. There may be texts in Peirce that suggest this line, and given that there can be several quasi-natural conclusions to a story, one might make the case that this is where the story naturally leads. It is my contention, however, that while Peirce does view truth and reality as tied to scientific inquiry, it is not in the suggested way but rather in one importantly different from it in several crucial respects.

a. Truth

Taking the growth of knowledge to be a fact of history, Peirce explicates it in terms of the self-corrective feature of scientific inquiry and feels that this gives us good reason to think of an ultimate convergence of this inquiry and good reason to define truth in terms of this convergence. Drawing on the model of a converging series he defines truth as the limit of that series: "The opinion which is fated to be ultimately agreed upon by all who investigate is what we mean by the truth" (5.407). Truth is to be understood in terms of that body of assertions which is authorized in the ultimate account of the world. It is a property of the propositional content of a belief, so it cannot be completely independent of thought but "it is independent of all that is arbitrary and individual in thought, and quite independent of how you or I or any number of men think" (8.12). As a post-Kantian he eschewes the idea that truth involves some kind of correspondence with a mind-independent reality, but he insists that it does not follow from this that truth must be subjective or relative. As he understands it, truth is not relative to person, time, or circumstance but is the absolute limit toward which inquiry tends.

This is not to say that anyone will ever be in a position to know she has the truth or even to say that truth will in fact be attained, but only to say that this is what we mean by the concept of truth:

> I do not say that it is infallibly true that there is any belief to which a person will come if he were to carry his inquiries far enough. I only say that that alone is what I call truth. I cannot infallibly know that there *is* any truth. (*LW*, 26)

If one insists on the notion of consensus, truth then is a matter of the consensus that *would* emerge if a properly conducted scientific inquiry were to continue indefinitely.

This being truth's definition, our own beliefs may be said to be true in virtue of their agreement with or correspondence to this ideal: "Truth is the concordance of an abstract statement with the ideal limit towards which endless investigation would tend to bring scientific belief" (5.565). What we mean when we say that our present beliefs are true is that we have reason to believe that they agree with or approximate those that would come to be held if cognitive inquiry were able to come to its natural term. The 'agreement' here is not between thought and a mind-independent reality but between present representations and those in the ultimate account: "That to which the representation should conform is something in the nature of a representation ... and utterly unlike a thing-in-itself" (5.553). On this account of truth, truth and reality are intrinsically rather than extrinsically related:

> To make a distinction between the true conception of a thing and the thing itself is ... only to regard one and the same thing from two different points of view; for the immediate object of thought in a true judgment *is* the reality. (8.16)

Reality, then, is defined in terms of the representative capacity of the ultimate beliefs and is conceived of as "the normal product of mental action and not the incognizable cause of it" (8.15). On this account both truth and reality are tied to inquiry but in a way that preserves the objective and absolute character of both.

b. Reality

The above quotations, insofar as they bear on Peirce's concept of reality, are ambiguous, and the ambiguity masks a substantial development in his thought from an early phenomenalist strain to his later view, which might be conceived along internal realist lines. He used the same term "realism" to describe his position throughout. For Peirce the contrast term for realism was always "nominalism," and as such his use of "realism" has an ontological rather than epistemic thrust. The reality of generals or laws is what he has in mind. From the epistemic side, the common thread running through his various formulations of realism is his insistence on the cognizability of reality and his rejection of the thing-in-itself. Even given these constants, however, the difference between his early and late realisms is considerable.

In an early discussion of the realism-nominalism debate he puts the matter thusly:

> A realist is simply one who knows no more recondite reality than that which is represented in a true representation. Since, therefore, the word "man" is true of something, that which "man" means is real. The nominalist must admit that man is truly applicable to something; but he believes that there is beneath this a thing-in-itself, an incognizable reality. His is the metaphysical figment. (5.312)

The point here is that there is no reality independent of the representation relation. The concept of reality initially emerges experientially in contrast to that of an illusion, and the distinction is between "an *ens* relative to private inward determinations, to the negations belonging to idiosyncrasy, and an *ens* such as would stand in the long run" (5.311). The nominalist insists on explaining this experiential difference in terms of the latter representation being caused or constrained by the real object, which is conceived of as a concrete particular having its features completely independent of mind. The realist, on the contrary, resists this postulation of a mind-independent reality, and explains the phenomenological distinction in terms of the fact that some of our representations are subsequently discovered to be idiosyncratic, whereas others survive the test of time and are enriched by successive informational input which eventuates in a settled representation. This explanation seems sufficient: "To assert that there are external things which can be known only as exerting a power on our sense is nothing different from asserting that there is a general *drift* in the history of human thought which will lead to a general agreement, one catholic consent; and any truth more perfect than this destined conclusion, any reality more absolute than what is thought in it, is a fiction of metaphysics" (8.12).

On this account, the realities would "not be the unknowable causes of sensation but the intelligible conceptions which are the last products of mental action which is set in motion by sensation" (8.13). And since these intelligible conceptions will inevitably have an element of generality, the general must be acknowledged to be at least as real as the concrete. In fact, what is called into question on this account is the reality of individuals (what Peirce calls "singulars"), inasmuch as they could never be exhaustively captured in any representations: "The absolute individual cannot only not be realized in sense or thought but it cannot exist properly speaking" (3.93). He goes on to refer to such absolute individuals as "ideals" and maintains that "they do not exist as such" (5.311) because "being at all is being in general" (5.349). Reality has as a necessary condition representability, so individuals as such are merely *entia rationis*.

The later version of Peirce's realism carries over the same general idea of reality as that which is the object of the true representation and is thus independent of the idiosyncrasies of particular representations; but this later realism has an accommodation for individuals as well. The accommodation is made by Peirce's distinction between "reality" and "existence":

> *Reality* means a certain kind of non-dependence on thought and so is a cognitionary character, while *existence* means reaction with the environment, and so is a dynamic character; accordingly the two meanings ... are clearly not the same. (5.503)

This distinction is drawn in the context of developing an account of what he calls "the modes of being", namely, 'possibility', 'existence', and 'reality'. Reality is that mode of being the characteristic of which is that things are what they are independently of any assertion about them (6.349), whereas existence is the mode of being which consists in the genuine dyadic relation of a strictly determinate individual with all other individuals (6.336). Reality has to do with intelligibility, whereas existence involves brute interaction.

On this later account there is a place both for the reality of the intelligible objects of inquiry and the existence of concrete determinate particulars. Both are what they are 'independently of any particular representations' but in quite different senses; realities are independent of particular representations whereas existents are the external reactive constraints on representation. Each has its distinct mode of being. Generals such as kinds, properties, relations, and laws are real but are not said to exist, while individuals insofar as they are causally operative are said to exist. The realm of objects is cognizable and representable, and as such is the term of inquiry. On the other hand, the individual as such can be pointed to but not represented; it can be experienced but not cognized:

> What we commonly designate by pointing at it or otherwise indicating it we assume to be singular. But so far as we can comprehend it, it will be found not to be so. We can only *indicate* the real universe; if we are asked to describe it, we can only say that it includes whatever there may be that really is. This is a universal, not a singular. (8.208)

It is important to note that on this revised account, although the existent individual cannot be cognized, it is not a thing-in-itself to which we have no access. If we insist on this Kantian language, then Peirce puts his point quite boldly: "We have direct experience of things in themselves; nothing can be more completely false than that we can experience only our own ideas" (6.95). Although we do not have a cognitive grasp of the individual as such, it is present to us in the oppositional element in our experience. Our experience is not of our ideas but of the world in all its dimensions through our ideas. Perception is obviously the key. On his account, the realm of concrete individuals is present to us in perception, a feature of which involves the brute facticity of the world forcing itself upon us and resisting some of our ways of conceptualizing it (1.324; 336). This categorization in terms of modes of being enables Peirce to think of reality in terms of the representative capacity of the intelligible products of inquiry and still see the inquiry process as constrained by factors completely independent of it.

This fundamental distinction provides conceptual space for a view of confirmation as a real independent monitor of the process of scientific theorizing. Confirmation (the inductive phase of inquiry) functions by means of predicted perceptual judgments, and on this account such perception put us in contact with an independent world of individuals that, far from being the product of our thought, forces itself upon us. There is an element of compulsion and resistance in our perceptual contact with the world that enables this kind of cognition to be an effective monitor on speculation. Moreover, since perception is both a starting point and a testing ground for rational speculation, scientific inquiry can be viewed as being moored at both beginning and end by something radically independent of the cognitive process itself.

The real is still identical with the cognizable; the real world is the world represented in the ultimate scientific account. The general ontological picture, however, is more complicated. It is an ontology of individual reacting existents understood as certain kinds of things related in certain ways and falling under certain laws. The world so understood is the real world, and the items, structures, features, and relations of that world are *real* in the only sense Peirce can identify for that term.

c. *Scientific Realism*

Given that for Peirce the very concepts of truth and reality are tied to scientific inquiry, some of our ways of posing the question of scientific realism have quasi-definitional answers. First, is science concerned with truth? Quite clearly, for Peirce, truth is the defining aim of science, and he heaped scorn on those who took a more 'pragmatic' view of science, no matter how noble the other aims to which they might make science subservient: "I must confess that I belong to that class of scallawags who propose with God's help to look truth in the face, whether doing so be conducive to the interests of society or not" (8.143). For Peirce the goal of scientific theorizing is truth and, although there is much by way of practice that is intrinsic to the achievement of this goal, to make truth subservient to any practical aims, whether they be religious, social, or educational, was to pervert the scientific spirit: "Now the two masters, theory and practice, you cannot serve; that perfect balance of attention which is requisite for observing the system of things is utterly lost if human desires intervene, and all the more the higher and holier those desires may be" (1.162).

Secondly, does science get at the real structures of things? Again Peirce's answer is in the definitional mode. There simply is no more recondite reality than what science gets at! Reality is what would be represented in the ultimate scientific account if the inquiry continues to fruition. Hence, in our most general way of thinking about the issue of scientific realism, Peirce is clearly a scientific realist. Nevertheless, there are issues to be raised even about this most general sense of scientific realism.

The real issues at stake with regard to this general scientific realism are the issues of progress in science and the convergence of scientific explanation. That Peirce viewed the history of science as progressive is relatively incontrovertible: "We hope that in the progress of science its errors will indefinitely diminish just as the error of 3.14159, the value given for π, will indefinitely diminish as the calculation is carried to more and more places of decimals" (5.565). The "hope" ranges only over the "indefinitely," not the cumulative-progressive picture. This mathematical analogy greatly simplifies and overstates the sense in which science can be viewed as cumulative, but Peirce surely held what we might call a 'cumulation thesis'. The later stages of scientific explanation build upon the earlier ones, correcting their deficiencies while preserving their valuable insights. That the decimal expansion example was seen as simplistic is clear from Peirce's claim that science "advances by leaps; and the impulse for each leap is either some new observational resource or some novel way of reasoning about the observations" (1.109). He invokes evolutionary models when interpreting the history of science, but on his models its evolution proceeds by leaps "not by insensible steps" (1.109). However, it is a real evolution involving a definite accumulation of knowledge, whereby the later stages are advances over the earlier and give us a better grasp of the real structures of things.

This 'cumulation thesis' does not entail but certainly suggests what we might call the 'convergence thesis'. Given the definitions of "truth" and "reality" articulated by Peirce, if there is going to be any truth or reality, then the process of scientific inquiry will have to converge on a single most adequate representation of the way things are. Even this way of stating the 'convergence thesis' masks an ambiguity between the convergence of science on a unique understanding of *each* feature of the world and the convergence of scientific understanding on a *unified* representation of all features of the world. What Peirce seems to have in mind here is the former, the weaker convergence thesis, but even this claim can fail in two different ways: first, if the process of scientific inquiry is for some reason cut short and does not come to term, a settled view will not be reached; secondly, one might envision the inquiry being allowed full development but either settling on two or more quite different views of the nature of things or perpetually oscillating between two or more incompatible views. Realism with regard to scientific explanation would appear to be undercut by either eventuality.

The first possibility does not really undercut Peirce's version of this general sense of scientific realism because his claim is not that such convergence *will* occur but only that truth and reality are to be understood in terms of the convergence that *would* occur if scientific inquiry were allowed to come to term.[7] The second possibility, that the appropriate kind of inquiry develop indefinitely yet not settle on a unique view, is more to the point. A final multiplicity of significantly different explanations would seem to undermine the claim of any to be the true depiction of the real. Peirce would agree, but his agreement would be accompanied by the claim that such a scenario does not express a real possibility. The scenario envisaged is that of a multiplicity of empirically equivalent but ontologically diverse theories. However, given Peirce's pragmatic theory of meaning, the scenario imagined would not be seen to threaten his scientific realism but rather simply make the point that there might be two different ways of expressing the same truth and describing the same reality. Ontological claims are not independent of empirical decidability. Peirce's attitude toward the stronger convergence thesis, however, is less clear. The unity of science ideal, involving the interrelation or reduction of the various parts of science into one seamless web, is a more complicated and seemingly open question.

These, however, are not the issues that Peirce is thinking of when considering the question of realism and science. For Peirce, the contrast term to "realism" is always "nominalism," and what he has uppermost in mind is the issue of the status of laws, dispositions, and modality. Most of his discussion of the issue focuses on the reality of laws and is directed against nominalists such as Pearson and Mill, who held that the so-called "laws of nature" were not real facts about the world but purely mental constructs functioning as résumés of observed regularities. Peirce saw this kind of nominalism as rendering both science and common sense unintelligible.

He views the issue as one with the medieval problem of universals, the relevant class of universals now being the laws of nature "discovered" by science. Like the nominalists of old, the new nominalists view such laws as mental fictions having no ontological import. On the contrary, Peirce's view is that scientific laws purport to be straightforwardly true and as such to describe real features of the world, and that unless we take them this way we will be unable to make sense out of our normal interactions with our environment. Peirce tries to make his point with an example. Holding up a stone he asks—how do I know the stone will drop when I let it go? I can predict its dropping with confidence, and it seems true to say that "I *do know* that the stone will drop, as a *fact*, as soon as I let go my hold; if I *truly know* anything, that which I know must be *real*" (5.94). His question is, how could I know this fact about the world? He thinks that the obvious answer is that I know it by prediction from the general law that all heavy objects fall to the earth when unsupported. If you grant that I *do know*, it can only be because this law provides a reliable basis for prediction in virtue of the fact that the law "corresponds to a reality" (5.96).

Both the nominalist and the realist grant the uniformity in question but have different understandings of it:

> With overwhelming uniformity, in out past experience direct and indirect, stones left free to fall have fallen. Thereupon two hypotheses only are open to us. Either
>
> 1. the uniformity with which those stones have fallen has been due to mere chance and affords no ground whatever, not the slightest, that the next stone that shall be let go will fall; or
>
> 2. the uniformity with which stones have fallen has been due to some *active general principle*, in which case it would be a strange coincidence that it would cease to act at the moment my prediction was based on it...
>
> Of course every sane man will adopt the latter hypothesis ... that general principles are really operative in nature. (5.100)

Peirce's point is that you have no right to believe the *truth* of the prediction unless you believe in the *reality* of the law in question. Of course, it's possible that the law cease to operate the moment you make your prediction, and it is possible that chance events happen; the point is that the only way we could know that the stone would fall is if we knew the law and knew that it described some feature of the real world.[8] That we do know such facts about our world and can make justified predictions of this kind is obvious to Peirce; to be skeptical about this is to be "blinded by theory" (5.96). On the basis of this fact of justified prediction he mounts his argument: "My argument to show that law is reality and not figment,—is in nature independently of any connivance of ours,—is that predictions are verified" (8.153).

The fact that laws really are a feature of our world is not an isolated claim for Peirce but is embedded in other reflections about the natural history of law and the presuppositions of law. With regard to the former, Peirce gets quite speculative and supposes that the laws of nature themselves are the result of an evolutionary process from a state of things in the infinitely distant past, in which there were no laws to a limit state of complete determination in the future. Accordingly, at any given time the laws of nature at the micro-level are such as to admit of modest departures from them in the form of chance occurrences. Puzzling about the reflexive problem that if the laws of nature are the results of evolution, this evolution itself must proceed according to some principle which will also have the nature of a law, Peirce concludes that this meta-law must be such as to develop itself. He finds a model for this in habit formation and concludes this line of speculation with the hypothesis "that the laws of the universe have been formed under a universal tendency of all things toward generalization and habit-taking" (7.515).

His further reflection on the presuppositions of law lead Peirce to a discussion of the reality of dispositions and modalities. This again involves a critique of nominalism including, in particular, some earlier formulations of his own thought which he later saw to be nominalistic. The case in point is the example of the diamond at the bottom of the sea. In his early discussion of the pragmatic maxim, Peirce had raised the question whether or not a diamond at the bottom of the sea which would never be in a position to be scratched in the relevant sense could be meaningfully said to be hard. Since the meanings of terms like "hard" and "soft" were to be explicated in terms of conditionals referring to future experiences, Peirce had there stated that there would be no falsity involved in ascribing either predicate to the diamond because "the question of what would occur under circumstances which do not actually arise is not a question of fact but only one of the most perspicuous arrangement of them" (5.403). This view Peirce later saw as too nominalist:

> I myself went too far in the direction of nominalism when I said that it was a mere question of convenience of speech whether we say that a diamond is hard when it is pressed upon or whether we say that it is soft until it is pressed upon. I *now* say that experiment will prove that the diamond is hard as a positive fact. That is, it is a real fact that it *would* resist pressure, which amounts to extreme scholastic realism. (8.207)

He saw as nominalistic the reductionistic thrust of the pragmatic maxim whereby dispositions were reduced to their concrete manifestations and possibilities reduced to actuality.[9]

His realistic reformulation of his view involves a commitment to what he calls "objective modality." Taking possibility as an example, he distinguishes two uses. First, some instances of our use of the concept of possibility are clearly subjective, such as cases where we use "possible" to express the fact that we don't know that a given proposition is false, e.g., "I think John is thirty but he may possibly be thirty-five." Secondly, there are other cases of possibility that we normally think of as objective, as when we have the thought, "It's possible for one to graduate in three years." The first case is one where there is a completely determinate fact of the matter and the variability is only on the side of our beliefs; the second case points a real "variableness in the world" (5.455). This is how we normally think about possibility and the other modalities, and Peirce urges that we not reduce the second way to the first.

With regard to the question of the diamond: the question is, *was* that diamond really hard when there was no actual determination of it to be so, or *is* this diamond hard even if there will be no actual determination of it to be so? Is its hardness a real fact in the world? Peirce's mature answer is an unequivocal yes, and this answer involves a commitment to the reality of irreducible dispositions and modalities. To identify a given object as a diamond is to conceive of it as a certain kind of thing having a definite structure and certain essential properties from which "hardness is believed to be inseparable" (5.457). He then concludes rhetorically: "Is it not a monstrous perversion of the word and the concept *real* to say that the accident of the non-arrival of the corundum prevented the hardness of the diamond from having the *reality* which it otherwise with little doubt would have had?" (5.457) The diamond in fact has a structure which grounds real dispositional properties which, whether manifested or not, involve real objective possibilities in the world.

Talking in other places about probability rather than possibility, Peirce defines the meanings of statements about probabilities in terms of the "would be's" ascribed to objects in the world:

> To say that a die has a "would be" is to say that it has a property, quite analogous to any habit that a man might have ... and just as it would be necessary in order to define a man's habit to describe how it would lead him to behave and upon what sort of occasion—albeit this statement would by no means imply that the habit *consists* in that action—so to define the die's "would be" it is necessary to say how it would lead the die to behave on the occasion that would bring out the full consequence of the "would be"; and this statement will not itself imply that the "would-be" of the die *consists* in such behavior. (2.664)

Dispositions have to be explicated in terms of behavior but this does not mean that they are reducible to behavior. The analysis recommended by the pragmatic maxim does not eliminate or reduce laws, dispositions, and powers but simply gives us a way of expressing our concepts of them so that we can tell real instances of them from fictions.

Peirce is quick to point out, moreover, that their nonreducibility should not be taken in any mysterious sense. To say that an object has a certain dispositional property is just to say that if it were to be exposed to any agency of a certain kind, a certain kind of experienciable result would ensue: "For the pragmaticist, it is precisely his position that nothing else than this can be so much as *meant* by saying that an object possesses a character; he is therefore obliged to subscribe to the doctrine of real Modality, including real Necessity and real Possibility" (6.457).

This account of the modalities obviously involves a heavy commitment to counterfactuals, and the development of Peirce's thought on the matter from its earlier to its later stage is paralleled by a development in his analysis of counterfactuals. As one might expect, in what he identified as his earlier nominalist period he proposed a simple analysis of counterfactuals in terms of material conditionals. His expressed view was that there is no real contradiction involved in two apparently incompatible counterfactual conditionals, inasmuch as since both antecedents are false, both conditionals are true: "There is no objection to a contradiction in what would result from a false supposition" (5.403). Possibilities are reducible to actual events.

His later, more robust recognition of the reality of possibility went hand in hand with his rejection of the simple truth-functional analysis of counterfactuals:

> I soon discovered upon critical analysis that it was absolutely necessary to insist upon and bring to the front the truth that a mere possibility may be quite real. That admitted, it can no longer be granted that every conditional proposition whose antecedent does not happen to be realized is true.... (4.581)

There must be some feature of the real world that makes some subjunctive conditionals true and others false; the conditional form must have some real counterpart in the world. Such conditionals "must be capable of being true, that is, of expressing whatever there be which is such as the proposition expresses, independently of being thought to be so in any judgment or being represented to be so in any other symbol of any man or men; but that amounts to saying that possibility is sometimes of a real kind" (5.453).[10]

In addition to the question of realism about scientific laws there is also the question of realism about scientific entities, and the answers to these questions don't always go hand in hand. Although Peirce's focus is clearly on the former, one can tease out of the expressed view the Peircean position on the latter. On his account the ultimate scientific theory is true and as such describes the structures of the real world; and this theory refers not only to certain laws of nature but also to the kinds of things these laws range over. Scientific theory essentially involves a classification of the kinds of things that there are, and for Peirce "a classification is true or false, and the generals to which it refers are either reals in one case or figments in the other" (5.453). Hence it seems that Peirce is committed to the reality of the entities referred to in the ultimate scientific account.

What exactly does this commitment amount to? How does a commitment to the reality of scientific entities bear on the status of the entities referred to in our common-sense view of the world around us? Does the truth of the scientific picture of the world imply the falsity of our common-sense picture? Although these are not the questions that occupied center stage for Peirce, I think his view as articulated does contain the resources to sketch an answer to them.

That scientific explanation involves the postulation of theoretical entities seems clear, and that Peirce's attitude toward these entities is basically realistic seems equally so:

> The things that any science discovers are beyond the reach of direct observation. We cannot see energy, nor the attraction of gravitation, nor the flying molecules of gases, nor the luminiferous ether, nor the forests of the carbonaceous era, nor the explosions in the nerve cells. It is only the premises of science, not its conclusions, which are directly observed. (6.2)

This is a motley list of theoretical entities, but of all of them Peirce uses the language of "discovery." It is a conclusion of science that if (and hopefully 'only if') we think of the world as involving such entities we can explain what we experience of it.

William James in the preface to his *Principles of Psychology* had taken a more positivistic line and had eschewed as metaphysical all attempts to *explain* our experience as in any way a product of "deeper-lying entities." Peirce's critical response was that physicists at any rate did not confine themselves to such a "strictly positivistic point of view":

> Students of heat are not deterred by the impossibility of directly observing molecules from considering and accepting the kinetical theory; students of light do not brand speculations on the luminiferous ether as metaphysical; and the substantiality of matter itself is called in question in the vortex theory, which is nevertheless considered as perfectly germane to physics. All these are "attempts to explain phenomenally given elements as products of deeper-lying entities." In fact, this phrase describes as well as loose language can the general character of scientific hypotheses. (8.60)

The point is that it is the nature of scientific explanation to explain the features of the world that are more directly available to us in terms of "deeper-lying entities" to which we don't have direct access. It is Peirce's view that we can regard these scientific hypotheses as real explanations only if we take their postulated entities as real features of the world. Moreover, this scientific picture of the world can be quite different from the familiar world of common sense: "Modern science, with its microscopes and telescopes, with its chemistry and electricity, and with its entirely new appliances of life, has put us into quite another world; almost so much so as if it had transported our race to another planet" (5.513). The question that remains is the crucial one: Does the scientific picture of the world constituted by the realistic "taking" of its postulated theoretical entities commit us to the falsity of our common-sense picture of the world, which seemingly involves quite different kinds of entities?

On my reading the Peircean answer is "no." The relation of the scientific conception of the world to the ordinary conception is not that of the true to the false but rather that of the precise to the vague. An appreciation of this answer involves a grasp of Peirce's view of concept formation and the status of both common sense and scientific concepts.

For Peirce our various conceptualizations of the world are not simply given but are the result of mental processes; and his suggestion is that these processes have the same formal features both in the case of the formation of normal perceptual beliefs and the construction of scientific hypotheses.[11] In both cases it is a matter of coming up with a simple predicate (whether it be "red," "fish," or "electron") that reduces the manifold of experience to some kind of unity and thus explains it and renders it intelligible:

> Every concept, doubtless, first arises when upon a strong, but more or less vague, sense of need is superinduced some involuntary experience of a suggestive nature.... With man these first concepts (first in the order of development, but emerging at all stages of mental life) take the form of conjectures, though they are by no means always recognized as such. Every concept, every general proposition of the great edifice of science, first came to us as a conjecture. (5.480)

The mental processes that generate all our conceptualizations of the world, from the most general and primitive to the most precise and sophisticated, are inferential in nature, and the specific form of inference germane to the generation moment is abduction: "Perceptual judgments are to be regarded as an extreme case of abductive inferences, from which they differ in being absolutely beyond criticism" (5.181). It is with these considerations in mind that Peirce talks of ordinary perceptual judgments and scientific theories as both being conjectural or hypothetical in nature.

This is not to deny that there are enormous differences between ordinary perceptual judgments such as recognizing the colors or spatial relations of the middle sized physical objects in our environment, and scientific accounts of the charge on an electron or the structure of the gene; but the difference is not one of logical form—the processes leading to both judgments have an inferential structure and are "logically exactly analogous" (5.109). It's just that the former 'inferences' are unconscious, indubitable, and as such not subject to criticism, whereas the latter are paradigmatically conscious, dubitable, and subject to constant criticism.

As hypothetical products of abductive processes both orders of concepts are obviously fallible, but there is a clear sense for Peirce in which our ordinary perceptual judgments are indubitable and reliable. They are indubitable because they are beyond direct criticism, being of the general nature of instincts forced upon us below the level of critical reflection; and they are quite reliable because they are founded "on the totality of everyday experience of many generations of multitudinous populations" (5.522). Of course, it does not follow from the fact that they are in this sense indubitable that they are infallible or incorrigible, because I may subsequently have other perceptual judgments similarly forced upon me incompatible with those earlier and "on that basis may infer that there must have been some error in the former reports" (6.141); nor does it follow from the fact that they are generally reliable that they are true. Nevertheless, they are indubitable and reliable and as such are the touchstones of inquiry.

The positive epistemic status of these perceptual judgments, however, is circumscribed by their primary role in guiding our ordinary interactions with our environment and their status as necessarily vague:

> The indubitable beliefs refer to a somewhat primitive mode of life ... yet as we develop degrees of self-control unknown to that man occasions of action arise in relation to which the original beliefs, if stretched to cover them, have no sufficient authority; in other words, we

> outgrow the applicability of instinct—not altogether, by any manner of means, but in our highest activities." (5.511)

These perceptual judgments, then, have what authority they have only with regard to those levels of interaction with our environment "that resemble those of a primitive mode of life" (5.445) and commensurate with their status as "invariably vague" (5.446). Their epistemic authority is diminished with regard to what Peirce calls our highest activities; and what he has in mind by "highest activities" is the sphere of scientific theorizing, the sphere of activity that opens up to us another world, the world of "microscopes and telescopes with its chemistry and electricity" (5.513).

But if the epistemic status of our ordinary perceptual judgments is circumscribed by their vagueness, that of our scientific judgments is circumscribed by their ideality. Scientific theorizing necessarily involves a dimension of idealization, and the appreciation of this should affect our commitment to scientific entities as so idealized.

> In all the explanatory sciences theories far more simple than the real facts are of the utmost service in enabling us to analyze the phenomena, and it may truly be said that physics could not possibly deal even with its relatively simple facts without such analytic procedure. Thus, the kinetical theory of gases, when first propounded, was obliged to assume that all the molecules were elastic spheres, which nobody could believe to be true. (7.96)

Given the human condition, 'to understand is to simplify,' and we should bear this in mind in forming our attitudes towards the conclusions of our sciences: "Owing to the necessity of making theories far more simple than the real facts, we are obliged to be cautious in accepting any extreme consequences of them, and also be on our guard against apparent refutations of them based upon such extreme consequences" (7.96). Given these caveats, it is reasonable to believe that our theories do give us access to how things are.

The general picture Peirce has in mind seems to me to be recoverable from several texts:

> Those vague beliefs that appear to be indubitable have the same sort of basis as scientific results have. That is to say, they rest on experience—on the total everyday experience of many generations of multitudinous populations. Such experience is worthless for distinctively scientific purposes, because it does not make the minute distinctions with which science is chiefly concerned; nor does it relate to the recondite subjects of science, although all science, without being aware of it, virtually supposes the truth of the vague results of uncontrolled thought upon such experiences, cannot help doing so, and would have to shut up shop if she should manage to escape accepting them. (5.522)

Ordinary perceptual judgments made in standard conditions are reliable guides to and through the features of the world around us. They are vague but sufficient unto our ordinary purposes: "No words are so well understood as vernacular words ... yet they are invariably vague; and of many of them it is true that, let the logician do his best to substitute precise equivalents in their places, still the vernacular words alone, for all their vagueness, answer the principal purposes" (6.494). Among these purposes are the identification of given subjects of scientific inquiry and the identification of ranges of phenomena that can confirm or falsify scientific theories. Scientific theorizing, however, goes beyond these vague identifications in terms of both discrimination and precision. In theorizing, a new level of analysis is attained and conceptual principles that we have good reason to think are reliable at the macro-level may not at all apply at the micro-level, e.g., "it is quite open to reasonable doubt whether the motions of electrons are confined to three dimensions" (5.445). Modern science with its instruments and theories transports us to "quite another world" (5.513).

Moreover, it does not follow from the fact that the more precise scientific account is true that the original perceptually guided conceptualization is false. It could also be true at its level, and, in fact, given the role of perception in the origination and confirmation of scientific theories, the success of science would seem to depend on the world having "the character that it can be explored and progressively understood by means of vague ideas, provided only that they may be made sufficiently precise; in this sense vagueness is as evident a character of the objective world as preciseness."[12] Peirce goes so far as to say that what we might call our ordinary perceptual conceptualization of the world has "within its proper sphere ... more weight than any scientific results" (5.522). And it is crucial for science that this be so. Scientific theorizing develops within our ordinary, perceptually guided interactions with the world in such a way that the epistemic statuses of both are inextricably intertwined.

4. Fallibilism, Probability, and Confidence

The view of belief that falls out of this account of explanation is a species of probabilism bounded on one side by a 'contrite fallibilism' about all specific claims and on the other by a robust confidence that truth will be attained in the end. The sense in which this is Peirce's general epistemological position will be discussed in the next two chapters, but here I want to develop these features of his view specifically with regard to science.

a. Fallibilism

The multi-dimensional character of Peirce's fallibilism is articulated in one of his more general statements:

> There are three things to which we can never hope to attain by reasoning, namely, absolute certainty, absolute exactitude and absolute universality. We cannot be absolutely certain that our conclusions are even approximately true.... We cannot pretend to be even approximately exact.... Finally, even if we could ascertain with

> absolute certainty and exactness that the ratio of sinful men to all men was 1 to 1; still among the infinite generations of men there would be room for any finite number of sinless men without violating the proportion. (1.141)[13]

Our scientific beliefs are secured to the degree that they are confirmed, and the logic of inductive procedures precludes the possibility of absolutely decisive confirmation. The very design of the procedure rules out absolutes both of scope and detail.

Fallibilism for Peirce is more than anything a matter of attitude. Speaking of science he asserts that "its accepted propositions are but opinions at most, and the whole list is provisional" (1.635), and of the scientific man he maintains that he "is not in the least wedded to his conclusions; he stands ready to abandon one or all as soon as experience opposes them" (1.635). In fact, then speaking more precisely, he distinguishes between belief and acceptance with these considerations in mind:

> Really the word belief is out of place in the vocabulary of science. If an engineer or other practical man takes a scientific result and makes it the basis for action, it is he who converts it into a belief. In pure science, it is merely the formula reached in the existing state of scientific progress. (7.185)

Belief for Peirce is a very practical matter: "What I believe is what I am prepared to act on *today*" (7.606). And for him science as such does not have these short-range concerns in mind. With regard to scientific conclusions Peirce prefers to think in terms of "modes of acceptance" (7.187).

Peirce's fallibilistic attitude is grounded in his firsthand experience as a working scientist. His own work ranged from the observational level to the purely theoretical, and it all confirmed him in his negative attitude toward certitude, exactitude, and universality: "Positive science can only rest on experience; and experience can never result in absolute certainty, exactitude, necessity or universality" (1.55).

A good percentage of his adult life was spent as an experimentalist. His energies were devoted to making measurements, designing instruments for making measurements, and drawing general conclusions from the measurements made, e.g., the precise longitudinal determinations of points on the globe and the value of the gravitational constant. Accordingly, he felt that his views on such matters should carry some authority:

> To one who is behind the scenes and knows that the most refined comparisons of masses, lengths and angles, far surpassing in precision all other measurements, yet fall behind the accuracy of bank accounts, and that the ordinary determinations of physical constants, such as appear from month to month in the journals, are about on a par with upholsterer's measurements of carpets and curtains, the idea of mathematical exactitude being demonstrated in the laboratory will appear simply ridiculous. There is a recognized method of estimating the probable magnitudes of errors in physics—the method of least squares. It is universally admitted that this method makes the errors smaller than they really are; yet even according to that theory an error infinitely small is infinitely improbable. (6.44)

One of his earliest scientific papers was "On the Theory of Errors of Observation,"[14] and since he knew he was working in one of the most precise corners of one of the most precise sciences, he felt confident in his generalization: "In those sciences of measurement which are the least subject to error—metrology, geodesy and metrical astronomy—no man of self-respect ever now states his result without affixing to it its *probable error*; and if this practice is not followed in other sciences it is because in those the probable errors are too vast to be estimated" (1.9). From his earliest days as a computer for the Coast Survey, Peirce was intimately familiar with the established way (the method of least squares) for determining the best value from a set of observations and thus determining what it is reasonable to believe to be *the* observational data. Obviously, for him, observation was no simple matter.

If "humanum est errare" is true even of the observational component of science, it is magnified when the various theoretical components are taken into account. The heuristics of abductive theory formation preclude exhaustiveness, and the very logic of confirmation precludes decisiveness. Reflecting on these features of scientific explanation, Peirce draws this conclusion:

> On the whole, then, we cannot in any way reach perfect certitude or exactitude. We can never be absolutely sure of anything, nor can we with any probability ascertain the exact value of any measure or general ratio.
>
> This is my conclusion after many years of study of the logic of science. (1.147)

Science teaches us that error is always and everywhere possible and that "our relation to the universe does not permit us to have any necessary knowledge of positive facts" (1.608).

These conclusions follow from the very logic of empirical explanation, but if we add to the mix ordinary human fallibility, in the sense of the possibility in practice of making mistakes, then Peirce extends his fallibilistic attitude even into the realm of the deductive sciences: "Mathematical certainty is not absolute certainty; for the greatest mathematicians sometimes blunder, and therefore it is possible—barely possible—that all have blundered every time they have added two and two" (4.478). While he does think that deductive reasoning is necessary and understands necessity in terms of truth in all possible universes (4.431), he points out that "when we talk of deductive reasoning as being necessary, we do not mean, of course, that it is infallible" (4.531).[15]

In addition to his reflections on the logic of science and general human finitude, there are metaphysical considerations that motivate Peirce's fallibilism. The world itself contains real elements of growth and spontaneity, making the quest for absolute certitude and exactness not only fruitless but inappropriate. Even at the macro-level, the more precise our attempts to verify any law the more certain we are to find irregular departures from law. The deep reason for this is that the deviation is "due to arbitrary determination or chance" (6.46). This being the case on the macro-level, when we move down to the micro-level Peirce thinks the case is even more obvious: "There is room for serious doubt whether the fundamental laws of mechanics hold good for single atoms" (6.11). Metaphysical indeterminacy seems clear; our laws are by nature statistical. There is real novelty and diversification continually taking place in the world, such that the events in the universe are law-like only to a degree of approximation. Moreover, "this degree of approximation will be a value for future scientific investigation to determine, but we have no more reason to think that the error of the ordinary statement is precisely zero than any one of an infinity of values in the neighborhood" (1.402).

Peirce in no way thinks that these three sets of considerations entail a skeptical stance. The positive side of his claims that "every proposition which we can be entitled to make about the real world must be an approximate one" (1.404) and that "the conclusions of science make no pretense of being more than probable" (6.39) is that it *does* make sense to think of the conclusions of science as approximately true and it *does* make sense to think of them as probable. The notion of approximate truth or likelihood is best understood in terms of his picture of convergence; and he devoted considerable attention to the notion of probability.

b. Probability and Likelihood

Not surprisingly, Peirce's views on probability underwent the same development as his pragmatic maxim, namely, from what he regarded as a nominalistic to a realistic version thereof. The early version is an instance of a frequency theory, whereas his later reflections fall into the category of a propensity theory; and the relations between the views are complex.

What both versions have in common is the rejection of the classical Laplacean conception of probability, which Peirce viewed as inappropriately subjective and a priori. His specific focus was the principle of insufficient reason, the classical notion that the state of ignorance is denoted by the probability .5, so that in the absence of information all events of which we are so ignorant are equiprobable. He argued that this feature of the classical view gave rise to inconsistencies:

> Let us suppose that we are totally ignorant what colored hair the inhabitants of Saturn have. Let us, then, take a color chart in which all possible colors are shown shading into one another by imperceptible degrees. In such a chart the relative areas occupied by different classes of colors are perfectly arbitrary. Let us enclose such an area with a closed line, and ask what is the chance on conceptualistic principles that the color of the hair of the inhabitants of Saturn falls within that area? The answer cannot be indeterminate because we must be in some state of belief; and, indeed, conceptualistic writers do not admit indeterminate probabilities. As there is no certainty in the matter, the answer lies between *zero* and *unity*. As no numerical value is afforded by the data, the number must be determined by the nature of the scale of probability itself, and not by calculation from the data. The answer can, therefore, only be one-half, since the judgment should neither favor nor oppose the hypothesis. What is true of this area is true of any other one; and it will equally be true of a third area which embraces the other two. But the probability of each of the smaller areas being one-half, that for the larger should be at least unity, which is absurd. (2.679)

Peirce saw the problem as a simple confusion of subjective and objective considerations. We are interested in a fact in the world, namely the color of hair. Given our ignorance of the matter we assign its having a given specific color a probability of .5. Of two possible colors each would have that probability, which should make the disjunction certain. Since there are any number of possible colors, the scenario is absurd.

Now there is something in this situation of ignorance that does have a probability of .5, but is is not that, for example, a given Saturnian has brown hair; it is something subjective: "There is a sense in which it is true that the probability of a perfectly unknown event is one half; namely, the assertion of its occurrence is the answer to a possible question answerable by 'yes' or 'no', and of all such questions just half the possible answers are true" (2.747). What's equiprobable is not the series of events in question, but the series of our answers 'yes' or 'no' when forced to choose one of the answers given the question—"will the event occur?" But this is a subjective matter; what Peirce wants to get a grip on is the notion of objective probability.

In his early, positive treatment of the question he was satisfied that a simple frequency theory constituted a sufficient account of the notion of probability. His version was formulated directly in terms of arguments and indirectly in terms of a series of events:

> Probability is a kind of relative number; namely, it is the ratio of the number of arguments of a certain genus which carry truth with them to the total number of arguments of that genus, and the rules for the calculation of probabilities are very easily derived from this consideration....
>
> To find the probability that from a given class of premises, *A*, a given class of conclusions, *B*, follows, it is simply necessary to ascertain what proportion of times in which premises of that class are true, the appropriate conclusion is also true. In other words, it is the number of cases of the occurrence of both events *A* and *B*, divided by the total number of cases of the occurrence of the event *A*. (2.657–58)

A specific argument of the form"A, therefore probably *B*" is to be treated as an instance of a series of arguments of this form.

The picture is that as we draw inference after inference the ratio of success settles down and begins to approximate a fixed limit. If we continue long enough, then, "in the long run there is a real fact which corresponds to the idea of probability, and it is that a given mode of inference sometimes proves successful and sometimes not, and that in a ratio ultimately fixed.... we may, therefore, define the probability of a mode of argument as the proportion of cases in which it carries truth with it" (2.650). In this way we can give meaning to the concept of the probability of *B* given A. The real fact corresponding to a given probability statement is an actual ratio.

On this early view probability is defined in terms of actual frequencies that converge on a limit. When Peirce later came to reflect on this analysis he called himself to task for so defining probability in terms of actual frequencies:

> When I come to define probability I repeatedly say that it is the quotient of the *number* of occurrences of the event divided by the *number* of occurrences of the occasion. Now this is manifestly wrong, for probability relates to the future; and how can I say how many times a given die will be thrown in the future?... For it is plain that, if probability be the ratio of the occurrences of the specific event to the occurrences of the generic occasion, it is that ratio that there *would be* in the long run, and has nothing to do with any supposed cessation of the occasion. (2.661)

Hence we are not talking about actual series but possible series, not actual frequencies but possible frequencies. The change in his view with regard to probability straightforwardly maps onto his earlier explored change from a nominalistic to a realistic view of reality and possibility.

The next step continues the parallel. The analysis in terms of frequencies—even possible frequencies—is judged not to be deep enough. The frequency itself is to be understood in terms of dispositions or propensities. The statement that a die has a certain probability of turning up six "means that the die has a certain 'would be'; and to say that a die has a 'would be' is to say that it has a property, quite analogous to any *habit* that a man might have" (2.663). Hence to talk about 'A being probable to a certain degree' is an elliptical way of talking about frequencies of success in classes of argument forms and indirectly about frequencies in classes of events, which frequencies are ultimately to be understood in terms of dispositional properties of certain natural kinds of entities. The disposition is explicated in terms of the frequency, but that does not at all mean that it is reducible to it: "To define the die's 'would be' it is necessary to say how it would lead the die to behave on an occasion that would bring out the full consequence of the 'would be'; and this statement will not of itself imply that the 'would be' of the die *consists* in such behavior" (2.664).

This strict notion of probability, however, does not apply to all scientific beliefs. In fact, as defined, it applies only to repeatable argument forms that are capable of being given a numerical ratio, and the most interesting scientific beliefs do not fall into this class. On his view of probability, it is meaningless to talk about the probability of theories or laws "as if we could pick universes out of a grab-bag and find in what proportion of them the law held good" (2.780). However, for these more interesting scientific beliefs, which rest on a more complex web of evidence, he does define other epistemic terms which capture different kinds of "imperfections of certitude" (2.262).

The first of these is *plausibility*, which Peirce uses with references to the epistemic status of theories at the abductive stage of inquiry prior to any confirmation:

> By plausible I mean that a theory that has not yet been subjected to any test, although more or less surprising phenomena have occurred which it would explain if it were true, is in itself of such a character as to recommend it for further examination, or, if it be *highly* plausible, justify us in seriously inclining toward belief in it, as long as the phenomena be inexplicable otherwise. (6.662)

A theory, then, is to be deemed plausible to the degree that it is recommended for further investigation by the heuristics of abduction, and it is highly plausible if it is the only theory so recommended.

The second notion in this neighborhood is *likelihood* or *verisimilitude*:

> I call the theory *likely* which is not yet proved but is supported by such evidence that if the rest of the conceivably possible evidence should turn out upon examination to be of a *similar* character, the theory would be conclusively proved. (2.663)

A theory is to be deemed likely if it is on the road to firm confirmation on the assumption that the relevant evidence yet to come in is of the same general character as that already accumulated. Peirce acknowledged that it would be a considerable advantage to be able to quantify the degree of likelihood a given theory had attained at a given time, but was not at all sanguine about the possibility of so doing: "Any numerical determination of likelihood is more than I can expect" (2.663).

The third notion is that of *practically certain* or *sufficiently proved*. Although only purely formal reasoning can result in demonstration or certainty, he felt that "the kind of reasoning which creates likelihoods by virtue of observations may render a likelihood *practically* certain" (2.664) and that certain theories were "sufficiently proved" (2.663) if, given the evidence, all views to the contrary were without any plausibility. Hence, far from holding that his fallibilism entailed skepticism, Peirce felt that there was a whole spectrum of positive epistemic statuses, including probability, plausibility, likelihood, and practical certainty, that were possessed by given scientific claims. Moreover, he held that there is a sense in which we can be absolutely confident that science will ultimately get at the truth.

c. *Confidence*

Peirce's most general attitude toward science is quite the antithesis of skepticism, both with regard to the appropriate scientific attitude and with regard to the outcome of scientific inquiry. He maintains that the scientific attitude "has never faltered in its confidence that it would ultimately find out the truth concerning any question in which it could apply the check of experiment" (7.87) and the much stronger thesis that "science is foredestined to reach the truth of every problem with as unerring an infallibility as the instincts of animals do their work" (7.77). Both of these claims have been explored in some detail in the discussion of the conditions of possibility of science. The first identifies one of the virtues necessary for the continuation of scientific inquiry, while the second claim is grounded in the self-corrective feature of scientific method together with the hope that the inquiry will continue indefinitely. They are again mentioned here only to provide closure for the discussion of fallibilism by showing how it is embedded in a wider view of science that he sometimes calls 'infallibilism'.

This confidence with regard to the attainability of the overall goal of science is counterbalanced by an attitude of great modesty with regard to the accomplishments of science to date:

> Persons who know science chiefly by its results—that is to say, have no acquaintance with it at all as a living inquiry—are apt to acquire the notion that the universe is now entirely explained in all its leading features; and that it is only here and there that the fabric of scientific knowledge betrays any rents.
>
> But in point of fact, notwithstanding all that has been discovered since Newton's time, his saying that we are little children picking up pretty pebbles on the beach while the whole ocean lies before us unexplored remains substantially as true as ever, and will do so though we shovel up the pebbles by steam shovels and carry them off in carloads. An infinitesimal ratio may be multiplied indefinitely and remain infinitesimal still. (1.116–17)

Peirce acknowledges science's radical incompleteness with regard to both scope and depth. On the one hand, he maintains that science to date has concerned itself with only a very limited number of the relations that actually obtain between the objects in our world, while, on the other hand, he adds that even given this restricted domain its results have been "altogether superficial and fragmentary" (1.119). So, while he does share the nineteenth century's optimism as it bears on the progress of science, he certainly does not share the view that the explanatory framework of science is for the most part complete.

5. Socio-Historical Conception of Science

Since human life is a socio-historical phenomenon, it should not be surprising that someone whose most fundamental characterization of science is as mode of life should conceive of science as essentially social and historical. One can construe the social and historical dimensions of science, however, as either shallow or deep, and it can be instructive to view Peirce's thought with this distinction in mind.

The shallow conception of the socio-historical dimensions of science is the simple acknowledgement that science as a human activity is embedded in social and historical networks that affect its functioning and development. Peirce was surely cognizant of the importance of these considerations. His own work in geodesy drove home to him the fact that the success of his investigations clearly depended on the information in the form of measurements that he received from other times and other places. Moreover, it also depended on the continued support of the U.S. Coast Survey, without whose funding and organizational network his various experimental projects would be doomed. These considerations are obvious; and the acknowledgement of these factors alone constitutes only a shallow view of the socio-historical character of science. It could be the case that the social and historical dimensions of the behavior of scientists and the various factors involved in the institutionalization of science were viewed as quite external to the logic of science and did not determine its content or circumscribe its rationality. For Peirce, however, the social and historical factors cut much deeper.

Social, historical, and valuational factors are viewed as built into the very logic of scientific explanation and development. First, the heuristic principles he invokes in his account of the *abductive* phase of inquiry are intrinsically historical and social. A theory is initially recommended not because of its likelihood of being true but in virtue of the role it can play at a given historical stage of the inquiry process; moreover, it is recommended not because of its place in the intellectual economy of the individual inquirer but of the community of inquirers. Secondly, the very logic of *induction* is similarly social and historical. The structure of confirmation is such that it simply makes no sense to think of it asocially or ahistorically. Peirce puts the point very starkly—if we are only interested in our own beliefs and in present inferences, we are doomed to being illogical:

> Logicality inexorably requires that our own interests shall not be limited. They must not stop at our own fate, but must embrace the whole community. This community, again, must not be limited, but must extend to all races of beings with whom we can come into immediate or mediate intellectual relation. It must reach, however vaguely, beyond this geological epoch, beyond all bounds. He who would not sacrifice his own soul to save the whole world, is, as it seems to me, illogical in all his inferences collectively. Logic is rooted in the social principle. (2.654)

History and community are built into the very logic of scientific inquiry, i.e., to understand science is to understand it as developmental and communal. Far from being compromised by these factors, scientific rationality is constituted through them.

Values and interests more generally are similarly intrinsic to scientific rationality: "The most vital factors in the method of modern science have not been the following of this or that logical prescription—although these have had their value too—but they have been the *moral factors*" (7.87). What grounds the distinctiveness of science as practiced and makes possible its continued development is a certain set of virtues being embodied in the individual members of the community of investigators. The set includes what we would call intellectual and moral virtues, e.g., those ranging from veracity to altruism, but for Peirce they are all epistemically relevant and constitutive of the concept of scientific rationality. From the perspective of the present, the account of science articulated by Peirce can be seen to embody features of the classical picture of science together with features emphasized by later historical and sociological understandings thereof. The philosophically interesting point is that these two sets of features, which from our vantage point are supposed discordant, seem to fit together in something like a coherent whole.

In the classical spirit he maintains (1) that the aim of science is objectivity and truth, (2) that there is a specifically characterizable scientific method, and (3) that this method defines the paradigmatic case of rational cognitive behavior. He also holds (4) that the history of science exhibits progress toward objectivity and truth, and (5) that there is a logic of scientific inquiry that provides a rationale for this progress. But he *also* goes on to insist that scientific inquiry is (6) informed by interests, (7) structured by norms, and (8) driven by certain ineliminable moral factors and social ideals. Added to these features are (9) a causal-evolutionary account of scientific insight and (10) a fundamentally economic characterization of theory acceptance.

The point most worth reflecting on is that for Peirce all these factors are 'intrinsic' to science and all are involved in an 'internal' account of scientific development. The last five do not compromise or even qualify the first five. In fact, far from being in conflict with any features of the classical picture, the social, historical, and valuational factors are seen to be partially constitutive of 'scientific rationality', 'scientific progress', and 'the realistic reach of science'.

Several features of Peirce's socio-historical conception of science come to the fore in his account of the role of *authority* in science. It will be recalled that the notion of authority was introduced as the characteristic mechanism in the second method of fixing belief (illustrated by the medieval period), which Peirce found to be seriously defective and with regard to which the scientific method was deemed to be cognitively superior. We should not infer from this, however, that he thought that authority had no cognitive role in science.

On the contrary, for Peirce authority plays a crucial epistemic role in the development and credibility of science:

> I do not hesitate to say that scientific men now think much more of authority than do metaphysicians; for in science a question is not regarded as settled or its solution as certain until all intelligent and informed doubt has ceased and all competent persons have come to a catholic agreement, whereas 50 metaphysicians each holding opinions that no one of the other 49 can admit, will nevertheless severally regard their 50 opposite opinions as more certain than that the sun will rise tomorrow. This is to have what seems an absurd disregard for other's opinions; the man of science attaches a positive value to the opinion of every man as competent as himself so that he cannot but have a doubt of a conclusion which he would adopt were it not that a competent man opposes it; but on the other hand, he will regard a sufficient divergence from the convictions of the great body of scientific men as tending of itself to argue incompetence and he will generally attach little weight to the opinions of men who have been long dead and were ignorant of much that has been since discovered which bears upon the question at hand. (*W*, 2.313–14)

The point here is that the relevant authority is that which is intrinsic to the specific cognitive endeavor. Certain theories become entrenched; certain scientific investigators earn credibility in a given domain. It would be cognitively irresponsible not to accord these instances of authority considerable deference. In fact, one of Peirce's maxims of scientific reasoning is "what is questioned by instructed persons is not certain" (*W*, 2.356). The operative word in the maxim is "instructed" because it is tied to the specific domain under investigation: "For Agassiz to attach no weight to the opinion of Darwin or for Darwin to attach no weight to that of Agassiz would show a narrow-mindedness most fatal to the sober investigation of truth" (*W*, 2.357).

His quarrel with the Schoolmen was not that they deferred to authority but that their authorities were outside the domain under investigation. They deferred to old books and prelates rather than to those who were conducting serious inquiry into a given domain with the best conceptual and physical tools available. Again, for Peirce, relevant and competent authority is viewed not as an extrinsic threat to the rationality of science but as intrinsic to and partially constitutive of it.

Since for Peirce science is not construed as an idiosyncratic or isolated cognitive domain, but rather as our paradigmatic case of knowing, this account of science is embedded in a larger picture of knowledge and justification which is being developed simultaneously. He saw as one of the major achievements of modern science the development of "a new conception of reasoning" (5.363), which could be expanded in the direction of a general theory of knowledge, with the more specifically scientific reflections and the more general reflections mutually informing each other so as to generate a comprehensive picture of human knowing.

III. *Peirce's Critique of Cartesian Epistemology*

When one thinks of Peirce's theory of knowledge, one thinks first of the sweeping critique of Cartesianism contained in his 1868 *Journal of Speculative Philosophy* papers on cognition. Since this set of papers is designed as a critique of Cartesian epistemology, it would be helpful first to characterize what he conceived as the classical Cartesian position and then to orient ourselves to Peirce's own view via his critique of several features of the classical view. As a first approximation, he viewed Cartesian epistemology as the conjunction of methodism, foundationalism, internalism, and more generally, individualism.

1. Cartesianism

The methodist-particularist alternative grows out of a consideration of the problem of the criterion. We want to know what we know (the *extent* of our knowledge) and how we are to decide whether we know (the *criterion* of knowledge). But it seems that we can get a grip on the first only if we are already in secure possession of the second, and we can gain access to the second only if we already have a sure grasp of instances of the first. There is an at least apparent circle here, with the non-skeptical responses being of two types: 'methodists' are those who begin with a criterion of knowledge and on the basis of it determine the extent of our knowledge, whereas 'particularists' are those who begin with clear instances of knowledge and on the basis of these fashion the criterion thereof.

Peirce views Descartes as a methodist. He sees him as beginning with the criterion of indubitability and using it to identify clear instances of things we know. The list of items he so identifies may be rather modest (the 'cogito'), but the strategy is clear nonetheless. Things then become more dialectical as he educes from this one instance a more expansive criterion ('clearness and distinctness'), but the basic structure of the approach is still defined by the initial universal doubt and the criterion of indubitability it invokes.

The foundationalist-coherentist alternative grows out of the spectre of a regress of justification. Most of the beliefs that we have are indirectly justified, i.e., justified in terms of their dependence upon or inferability from others of our justified beliefs. It seems, however, that if any of our beliefs are going to be ultimately justified not all can be justified in this way; some have to be directly justified and thereby capable of functioning as ultimate sources of the justification of our other beliefs. These directly justified beliefs are thought of as *basic* and variously characterized as self-evident, incorrigible, or at least self-authenticating in some weaker sense. The justification of our other beliefs rests on this foundation and is in one or another sense *derived*. The coherentist rejects this picture of justification as being both impossible and unnecessary. He feels that the various candidates for directly justified beliefs cannot stand up to criticism and that the alternative is not a simple looping of individually justified beliefs doubling back upon themselves but rather a holistic picture of justification wherein individual beliefs have their justification only derivatively. Our various beliefs are justified to the degree that they form part of a coherent set of beliefs that mutually support each other and together can function as effective guides for action. Justification comes from mutual interdependence, not from firm foundations.

Peirce sees Descartes as clearly a foundationalist. Taking an axiomatic system as a model, Descartes is looking for some small set of beliefs about which he cannot possibly be mistaken, which can then function as axioms for generating other justified beliefs about the world. The basic beliefs are thought to be indubitable, true, self-evident, and thus directly justified; moreover, this original justification can be logically communicated to our other beliefs about the world.

The internalist-externalist alternative is not quite so straightforward. This issue concerns what can confer justification on a belief, what can transform true belief into knowledge. Invoking a notion of 'direct access' where the contrast notion is inferential access (as opposed to a contrast between unmediated/mediated access), the strong internalist claims that the only real justifiers are items to which the subject has direct access and that the only items that meet this condition are the mental states of that subject. So conceived, internalism involves a two-part claim: that we do have direct access to our own internal states, and that justification is to be understood ultimately in terms of these states. The externalist generally operates with a different picture of human cognition. For him, what we directly know are not our internal states but rather objects in the external world; moreover, whether or not we are justified in our beliefs about them has more to do with the reliability of our belief-forming mechanisms and/or their causal dependence on the relevant extrinsic objects than on anything available to us in our own internal mental repertoire.

Peirce views Descartes as in the internalist tradition. For Descartes, we have direct access only to our own internal states, so it is on these ultimately that our epistemic justification rightly rests. It is true that once we have established the existence of a non-deceiving God we then have grounds for acknowledging the reliability of our cognitive faculties; nonetheless this derivative external means of justification is ultimately grounded not in the mere existence of the non-deceiving God but in our grasp of the existence thereof. What looks like an externalist feature of justification is reducibly internalist.

The individualist-socialist alternative grows out of a consideration of the fundamental locus of knowledge. The individualist views the individual knower as epistemically autonomous, i.e., the source and repository of justification and truth, and not epistemically beholden to, or threatened by, the agreement or disagreement of fellow inquirers. On the contrary, one inclined toward a social theory of knowledge will view the phenomena of agreement and disagreement as intrinsic to the cognitive status of his own beliefs. In some sense the community of relevant inquirers is the fundamental locus of knowledge.

Peirce saw Descartes as clearly an epistemic individualist. The Cartesian "teaches that the ultimate test of certainty is to be found in the individual consciousness" (5.264) and also "makes single individuals absolute judges of truth" (5.265). Peirce caricatures the Cartesian position as ultimately reducing to the stark claim that "whatever I am clearly convinced of, is true" (5.265). This epistemic individualism, together with the other three characteristic features of Cartesianism, he subjects to radical critique.

2. Critique of Cartesianism

a. The Critique of Methodism

The very idea of starting with a universal doubt and the criterion of indubitability strikes Peirce as both impossible and self-deceptive. The project of wiping the cognitive slate clean, so as to make possible a presuppositionless inquiry, he sees as the occasion for mere pretense:

> We cannot begin with complete doubt. We must begin with all the prejudices which we actually have when we enter upon the study of philosophy. These prejudices are not to be dispelled by a maxim, for they are things which it does not occur to us can be questioned. Hence, this initial skepticism will be mere self-deception, and not real doubt.... A person may, it is true, in the course of his studies, find reason to doubt what he began by believing; but in that case he doubts because he has a positive reason for it, and not on account of the Cartesian maxim. Let us not pretend to doubt in philosophy what we do not doubt in our hearts. (2.265)

The kind of radicality proposed by the Cartesian strategy is just pretense; the doubt it invokes is just a "counterfeit paper doubt" (6.498), and the pure inquiry it envisages is "no more wholesome than any other humbug" (2.147). Moreover, the whole Cartesian project, conceived as the effort to root out all presuppositions, seems to Peirce naive. The most crucial of our presuppositions are probably not available to present consciousness but come to light only given time and distance. It is often the beliefs men *betray* and not those which they *parade* which are decisive (5.445, n. 1).

Accordingly, the extreme criterion of indubitability the Cartesian wields is singularly inappropriate. In fact, no beliefs would survive such a test and those which historically have been deemed to do so do not survive the succeeding judgments of history: "All those which repose heavily upon an 'inconceivability of the opposite' have proved particularly fragile and short-lived" (5.376, n. 1) and "what has been indubitable one day has often been proved on the morrow to be false" (5.514). Peirce himself has no quarrel with a more modest criterion of indubitability, which would mandate starting with "propositions perfectly free from all actual doubt; if the premises are not in fact doubted at all, they cannot be more satisfactory than they are" (5.376).

There is no intrinsically privileged starting point and, in particular, no given criterion in terms of which to sort out true beliefs from the false. One simply starts any inquiry with those particular beliefs one has no positive reason to call into question:

> There is but one state of mind from which you can "set out," namely, the very state of mind in which you actually find yourself at the time you do "set out"—a state in which you are laden with an immense mass of cognition already formed, of which you cannot divest yourself if you would; and who knows whether, if you could, you would not have made all knowledge impossible to yourself? (5.416)

We always embark on cognitive projects *in medias res*, with certain issues established and other issues in question. Doubt functions only against a background of belief, so that particular beliefs that we regard as true are the presuppositions of any given inquiry. Of course, they may eventually be discovered not to be true, but this realization is just the realization of human fallibility and it alone doesn't undermine the inquiry process, as long as self-criticism and boot-strapping are built-in features of the inquiry.

The particular claims to knowledge with which inquiry begins are quite varied but they can be seen to fall into several general types. Ordinary perceptual judgments made under standard conditions by hitherto reliable subjects are such as to command acceptance. If we aren't aware of any specific corruption of the information processing channels, it is reasonable to believe that we do in these standard circumstances grasp certain facts about the objects around us, "although we can never be absolutely certain of doing so in any special case" (5.311). He thinks this is also the case with regard to some obvious facts of history, such as the claim that Napoleon lived in the nineteenth century. The logical structure of both perceptual judgments and historical claims is that of an hypothesis, "yet practically they are perfectly certain; [and] … it would be downright insanity to entertain a doubt about Napoleon's existence" (5.589). In addition, there is a whole host of general beliefs of common sense such as 'fire burns' and 'all men are mortal', some basic forms of inference such as modus ponens and modus tolens, and some established facts of science that it would be both impossible and foolhardy to try to subject to serious doubt. These are all paradigm cases of knowledge in terms of which our various inquiries must take their moorings, even the inquiry into the methods of inquiry itself.

b. The Critique of Foundationalism

Peirce's critique of Cartesian methodism leads into his critique of foundationalism. Although one can view the motivation of the methodist as that of securing firm foundations for one's cognitive enterprise, it is usually particularist claims that most readily come to mind when one thinks of foundationalism. The basic move is to identify some particular beliefs which are self-authenticating and which can thereby serve to block the regress of justification; it may or may not be the case that the means of identifying them is via some general criterion. Peirce's critique of foundationism focuses on the claims made for such self-authenticating instances of belief, no matter how arrived at.

Peirce's conception of foundationalism and consequently his critique thereof focuses on the notion of *intuition*. He calls those allegedly self-authenticating, non-inferential knowings which make possible a foundationalist picture of knowledge "intuitions" and variously defines an intuition as "an absolutely first cognition," or as "a cognition not determined by a previous cognition of the same object and therefore so determined by something out of consciousness," or as "a premise not itself a conclusion" (5.213). The first two definitions pick out all cognitions with the relevant feature, whereas the third specifically focuses on those cognitions which are judgments, although this distinction between class and sub-class turns out to be not as sharp as one might think. The fundamental contrast term is "discursive cognition," with intuitive cognitions alleged to be those whose warrant is not derived from relations to other cognitions but is either purely intrinsic to them or is grounded in some cognitively unmediated relation to the world. Peirce argues that there are no such foundational cognitive items.

His critique of cognitive intuitionism has a broader target than Cartesian foundationalism. Cartesianism is formulated in terms of the foundations of knowledge being basic beliefs, i.e., beliefs that have their justification independent of their relations to other beliefs. A different, though structurally isomorphic view, would locate the foundations of knowledge not in basic beliefs but in some other mental states which have their warrant not in terms of their relations to other mental states but immediately in terms of their causal relations to the external object. Both versions have the cognitively relevant foundation being epistemically independent of other cognitive states. Peirce's use of "intuition" ranges over both versions of foundations, and his critique of foundationalism is similarly broad.

It is also clear that Peirce's principal target is a very strong version of foundationalism which characterizes the basic beliefs or cognitive states as indubitable, incorrigible, and/or self-evident. He is not arguing against the weaker thesis that some of our beliefs or states have *some* of their warrant directly, i.e., independent of their relations to other beliefs or states. Such a weak foundationalism, as opposed to pure coherentism, he in fact finds congenial. What he objects to is the view that there are some beliefs or states which are for us epistemically autonomous and function as epistemically unrevisable building blocks of our cognitive edifice.

In his general critique Peirce construes the foundationalist picture of knowledge as involving three distinguishable claims: first, that there are such cognitively basic items as characterized above; secondly, that we can infallibly identify them as such, so that they can function to ground our other cognitive claims; and thirdly, that without some such set of cognitively privileged items our knowledge would have no sure foundation and skepticism would ensue. Peirce challenges all three claims, each of which he deems necessary to the foundationalist position.

He first calls into question the second claim, namely, on the assumption that there are such self-authenticating instances of knowledge, the claim that we can infallibly identify them as such, so as to enable them to ground our other knowledge claims. Peirce's response is that even if it were the case that we have 'intuitions', we have no intuitive power of distinguishing an intuition from another cognition. True to his own methodological principles of trusting "rather to the multitude and variety of arguments than to the conclusiveness of any one" and making your argumentative case not in the form of "a chain which is no stronger than its weakest link, but a cable whose fibers may be ever so slender provided they are sufficiently numerous and intimately connected" (5.265), Peirce puts forward a whole battery of arguments against this facet of foundationalism.

His arguments can be roughly categorized as either, first, broadly historical; secondly, empirical; or thirdly, conceptual. First, he chooses to confront this second claim of foundationalism with the relevant historical record. If we did possess such a power of intuitively distinguishing our intuitions from our other cognitions, surely there would be reasonable agreement among men, or at least among philosophers, as to which cognitions are intuitive. History, however, attests to the radical disagreement among even those philosophers who recognize intuitions as to which cognitions are intuitive. History seems to suggest, then, that even if we have intuitions, we have no direct power of distinguishing them from our other cognitions (5.215).

In a more specifically empirical vein, Peirce appeals to the general phenomena of perceptual supplementation, the specifics of spatial and tonal perception, and engages in some speculative ruminations in the area of child psychology. On the first point, he notes how difficult it is in the concrete to distinguish between what we have seen and what we have inferred, these being the relevant differences between the intuitive and the discursive. Describing the accident-witness situation and the magician-audience situation specifically, he argues that "it is not always easy to distinguish between a premiss and a conclusion, that we have no infallible power of doing so, and in fact our only security in difficult cases is in some signs from which we can infer that a given fact must have been seen or must have been inferred" (5.216). Consequently, even were there a pure seeing independent of inferential supplementation, it seems clear that we can not infallibly identify instances of such. He specifies this point further by attending to the phenomena of spatial and tonal perception.

His argument vis-à-vis spatial perception is actually a nest of sub-arguments. First he points out that prior to Berkeley's study of vision, it had generally been believed that the third dimension of space was immediately intuited, but that now nearly everyone admits that a three-dimensional image is an interpretation of a two-dimensional impression (5.219). Secondly, he reminds the reader of the blind spot on the retina and argues that the space we immediately see (when one eye is closed) is not, as we had imagined, a continuous oval but a ring the filling out of which must be the work of intellectual supplementation (5.220). How could mere contemplation distinguish intuitional data from inferred projection in this case? Finally, he argues that even the perception of two dimensional space involves intellectual supplementation of which we are not aware (5.223). This perception appears to be an immediate intuition, but it can't be. If we were to see an extended surface immediately, our retinas would have to be spread out in an extended surface. Since, however, our retinas consist of discrete nerve points, our given information is not of a continuous surface but of a collection of spots, and the resultant perception of the two-dimensional continuous surface is a function of the reduction of these phenomena to unity via the conception of extension. There is, as a consequence, no way in which what is simply given here (on the assumption that something is given) is open to intuitive grasp.

He makes a similar point with regard to tonal perception (5.222). With respect to the pitch of a tone which seems to be simple and immediately perceived, he notes its dependence on the frequency of the vibrations which reach the ear and therefore, upon the rapidity with which certain impressions are brought about in the mind. Since these impressions must exist prior to the given tone, the sensation of pitch is determined by previous cognition; and this could never be uncovered by the mere contemplation of that feeling.

Peirce's final, even more general empirical consideration is little more than a suggestion in the area of child psychology (5.218). He suggests that it seems reasonable to believe that children possess all the distinct cognitive powers of an adult. If there were, then, an intuitive power of identifying intuitions, it would seem reasonable to believe that children would both have it and manifest it. Yet the child seems to know very little indeed about *how* he knows what he does know, being unable to distinguish between obvious cases of learning (e.g., his grasp of the English language) and alleged intuitions. Hence, he at least does not appear to be in possession of a faculty of distinguishing, 'merely by looking', between an intuition and a cognition determined by other cognitions.[1]

On the purely conceptual level, Peirce provides two brief argument sketches against the second thesis of foundationalism. First, he argues that the only evidence for this faculty seems to be our *feeling* that we can clearly distinguish our intuitions from our other cognitions (5.214). But, he goes on, the significance of this testimony depends entirely upon our having the power of distinguishing in this feeling whether the feeling itself is the result of education, old associations, or is itself an intuitive cognition. It presupposes, in short, that to which it allegedly testifies. Furthermore, once this question is raised, one seems faced with a disturbing regress: is this feeling infallible, and the judgment concerning it infallible, ad infinitum. Circularity and infinite regress constitute imposing conceptual hurdles.

Secondly, he suggests an argument from dreaming (5.217). A dream, so far as its content goes, is exactly like an actual perceptual experience and can be mistaken for one. But it seems clear that every representation in a dream is determined according to the laws of association by previous cognition. Accordingly, if there were intuitions in the make-up of our ordinary experience and we had an intuitive faculty of distinguishing intuitions from non-intuitions, it would seem that we would never confuse dreams with reality. But we clearly can make this mistake; moreover, when we do upon awaking distinguish dreams from reality, it is not on the basis of some direct intuition but more on the basis of what might be classified as circumstantial evidence, i.e., certain discoverable characteristics of ordinary experience that are usually lacking in dreams.

These are the kinds of considerations Peirce brings to bear against the second tenet of foundationalism. Before commenting on the force of these considerations, it is important to consider what Peirce thinks he will have achieved if this attack on the second tenet of foundationalism is deemed successful. He will have accomplished two things: first, given that each of the three claims is necessary to foundationalism, he will have made a substantive point that effectively undermines the foundationalist position; secondly, he will have made a methodological point that goes far toward undermining the claim that there even are intuitive cognitions in the first place. Inasmuch as it must now be agreed that no general faculty for identifying intuitions exists, the claim that there are particular intuitions has to be argued for under the normal canons of parsimonious explanation. The existence of intuitions in particular cases can be posited only if the phenomenon of human cognition cannot be explained without them, the preference methodologically being given to those other explanations invoking only known faculties acting under conditions known to exist (5.226).

This, however, is merely hypothetical. The adequacy of Peirce's arguments against this second thesis of foundationalism is the crucial issue. Since his arguments are so many and various, and since there is no single point on which they all hinge, any adequate evaluation of them would have to be correspondingly piecemeal. In spite of this, however, a few general comments can be made.

The general historical argument would have considerable force for someone (such as Peirce) whose conception of explanation and justification is intrinsically social and historical, but is not likely to be effective against those (like Descartes) who are likely to defend a strong version of foundationalism in the first place. The recognition of those cognitions which are really intuitions, and thus properly basic, is likely to be reserved to those individuals whose cognitive faculties are functioning properly, and such individuals are not likely even to be embarrassed by the notion of proper functioning being evidenced by (if not even partially defined by) the very recognition of the appropriate intuitions.

The empirical arguments are similarly ambiguous. For a philosopher like Peirce who does not invoke a sharp dichotomy between science and philosophy, such empirical considerations are obviously very weighty. Most foundationalists, however, are among those who think that philosophical considerations are in principle different from empirical or scientific considerations, and since their foundationalism is a philosophical position, they are not disposed to feel as threatened by these kinds of empirical arguments. This leaves the conceptual arguments alone as uncontroversially telling against the foundationalist position.

Although there is something appealingly ironic about the employment of 'a dream argument' against Cartesianism, it's not clear that it is decisive. For it could be the case that, although we cannot *always* intuitively distinguish intuitions from other cognitions, we *sometimes* can, and that this circumscribed ability would be sufficient to identify foundational cognitions. Of course, it might seem that we have no good reason to posit the existence of such a circumscribed faculty, but the dream argument by itself would not present an insurmountable conceptual obstacle to the move.

The second conceptual argument focused on the evidence for the faculty of intuitively recognizing intuitions consisting in our *feeling* that we have this fundamental ability to distinguish directly our intuitions from our other cognition. The challenge is to become clear about the epistemic credentials of this feeling. Is the feeling infallible and, if so, is our judgment to that effect infallible also? One might try to make the case that this kind of regress is benign by distinguishing attempts to justify the belief that the feeling functions infallibly from attempts to ground an infallible identification; but any straightforward attempt to ground inferentially the reliability of the feeling would seem to threaten the infallibility of the original identification of the cognitions about which it is alleged. In summary, given the number and variety of considerations brought forward, it would seem at the very least that the claim that we have a general faculty for the immediate recognition of intuitions is under considerable pressure.

Having accomplished this much, Peirce then turns to the first claim of foundationalism (that there are intuitions) and embarks on a critical analysis of the two strongest claimants for the status of intuitive cognition, namely, first, an intuitive knowledge of the self; and, secondly, an intuitive grasp of the different subjective characters of our own mental states. With regard to the self, it is clear that I know that I exist. The question is *how* do I know it—intuitively or as determined by previous cognition? Given his having called into question the second tenet of the foundationalist position, it can no longer be claimed as self-evident that we have such an intuitive self-consciousness, so positing it will have to be argued for in the conventional manner. To block this move, Peirce points to empirical facts that count against the hypothesis of intuitive self-knowledge, puts forward a simpler hypothesis of his own to account for self-knowledge, and finally defuses a conceptual argument to the effect that there must be such an intuition.

He first points out that rather than simply being there full-blown from the beginning, self-consciousness seems to be something that develops over time in small children. The relatively late emergence of the use of the pronoun "I" seems to be a bit of empirical evidence to this effect. Working from this starting point, Peirce constructs at a very modest level a contrary hypothesis of his own to explain the origin and nature of self-consciousness (5.228–36). On his account, knowledge is originally and primarily directed toward external objects. When a sound is heard by a child, he thinks not of himself as hearing but of the object as sounding; and when he wills to move an object, he thinks not of himself as willing but of the object as fit to be moved. Knowledge of the self arises only in and by means of social interaction; specifically, the self is an hypothesis to account for (in the sense of provide a locus for) ignorance and error. As the child becomes aware of ignorance (testimony of others later confirmed by his own experience), it is necessary to suppose a self in which this ignorance can inhere. Moreover, the phenomenon of error (past beliefs corrected in view of present experience and/or the testimony of others) can be explained on the supposition of a self which is fallible. On this account, knowledge of the self is always in fact inferential, but the inferences have become so habitual to us that the knowledge has the appearance of being an immediate intuition. What Peirce feels recommends this explanation is that it accounts for all the facts while positing only known faculties operating under conditions known to exist.

Peirce finally takes up what appears to him to be the most impressive conceptual argument for intuitive self-consciousness. It has been argued that since we are more certain of our own existence than of any other fact, and since a premise cannot determine a conclusion to be more certain than it is itself, it follows that our own existence cannot have been inferred from any other fact. Peirce's answer is simple and devastating. That a conclusion cannot be more certain than the sum of facts which support it is true, but it may easily be more certain than any one of those facts. In this specific case, "to the developed mind of man his own existence is supported by *every other fact*, and is, therefore, incomparably more certain than any one of these facts; but it cannot be said to be more certain than that there is another fact, since there is no doubt perceptible in either case" (5.237). As a consequence of both the felt adequacy of his own hypothesis and his counterarguments against the other hypothesis, Peirce concludes that there is no necessity for supposing an intuitive self-consciousness since the phenomena in question can easily be accounted for in other ways.

Considering the second presumptive case of alleged intuitive knowledge, namely, our ability to distinguish between the different kinds of our mental acts or states, Peirce disengages the arguments for such an intuitive power by offering a more plausible contrary hypothesis of his own (5.238–43). The strongest argument for such an intuitive ability is conceptual. If we had no such direct ability to distinguish, for example, between what we believe and what we merely conceive, we could never confidently distinguish them; for, if we were to try to do so by reasoning, the question would then arise whether the argument itself was believed or merely conceived, and this would require answering before the original argument would have any force. A regress would clearly ensue. Secondly, we have the immense phenomenological difference between seeing a color and imagining it, a difference that seems to be most readily explained by a direct identification of one kind of state as different from the other. Peirce clearly agrees that we can and do readily and reliably distinguish between these mental states; the point at issue is whether or not this ability demands the postulation of an intuitive power or whether the facts can be explained without this supposition.

With regard to the second case, the distinction between perceiving and imagining, Peirce questions the severance of these mental acts from their objects, because his view is that the immense difference between the immediate *objects* of sense and imagination sufficiently accounts for our ability to distinguish these kinds of mental acts readily but not immediately. We differentiate these acts or states on his view not by direct intuition but by an inference from certain observed characteristics of their objects. The distinguishing of belief from mere conception is a little more complicated. Peirce maintains that we distinguish a belief from a mere conception in most cases by means of a peculiar feeling of conviction, and the distinction can be understood in two ways: belief can be understood 'sensationally', as a judgment which is accompanied by this feeling, or 'actively', as that judgment on which an individual is disposed to act. Taking belief sensationally, the power of recognizing it amounts simply to the capacity for the sensation which accompanies the judgment; taking it actively, its identification can be inferred from our readiness to act. In neither case need we resort to intuition in order to explain our ability to draw the distinction.

Having handled these two alleged cases of intuitive knowledge, Peirce next embarks on a discussion of an issue 'in the neighborhood' which will reappear again in his critique of Cartesian internalism. One might back off the claim that our grasp of our mental states is strictly intuitive but insist instead that it is 'introspective', where this weaker notion does not involve a total lack of mediation but only the absence of a particular kind of mediation: "By introspection I mean a direct perception of the internal world ... a knowledge of the internal world not derived from external observation" (5.244). If we were to have this cognitive ability, our grasp of our own mental states could be thought to have sufficient epistemic privilege to function as the sole appropriate foundation of our cognitive edifice.

Peirce does not think we are capable of introspection in this sense. In view of his previous argument against our having an intuitive ability to distinguish between different kinds of mental states, it can no longer be claimed that the existence of such introspection is self-evident but only that the postulation of such a power is necessary to account for the facts. On the contrary, he thinks that the positing of such a power is not only unnecessary but also implausible given a careful analysis of human experience.

There is, of course, a sense in which any given experience has an internal object if all one means by this is that it is influenced by internal conditions and has a subjective dimension. Peirce is not concerned to deny this. What he is concerned to deny is that we have *knowledge* of our own mental states that is in no way parasitic on our knowledge of external objects and is in virtue of this fact epistemically privileged vis-à-vis that knowledge of public objects. On the contrary, his view is that my grasp of the fact that "I am experiencing redness" or that "I am experiencing disgust" essentially involves a grasp of prior notions such as "that car is red" or "that behavior is disgusting." His view is that our direct cognitive response is to characters of public objects and that it is a mark of "returning reason" to grasp features of our own mental states (5.247). Moreover, not only is it implausible to think that our knowledge of our own mental states is either intuitive or introspective, on his view these conscious mental states themselves are complexes synthesized from prior states of which we are unaware. But more of this later.

The claim for specific kinds of intuitive knowledge having been so challenged, one might surmise that the case for 'intuitionism', i.e., for absolute foundations, had been put to rest. Peirce, however, attributes one final line of defense to the intuitionist. Although he may not be able to provide defensible specific instances of intuition, he could, nevertheless, claim to have a transcendental argument to the effect that there must be such intuitive knowledge. Peirce formulates the argument in the following form: "Since we are in possession of cognitions which are all determined by previous ones, and these by cognitions earlier still, there must have been a *first* in this series or else our state of cognition at any time is completely determined according to logical laws by our state at any previous time" (5.259).

The "or else" clause indicates what the concern is. The spectre that haunts the foundationalist seems to be the picture of a belief system, when all beliefs are determined by prior beliefs, being one which is in no way constrained by a mind-independent world. So, if there is going to be such a constraint by the world, our series of beliefs must go back to a first which is first because it is not constrained merely by other beliefs but directly by the real external object. Put another way, our present beliefs may rest on prior beliefs and these on others prior still; but since any individual knower goes back only a finite time, there must be some belief not derived from prior ones in order for our beliefs about the real world to have begun. And if intuitions are necessary in this case, they are surely possible in others.

Peirce's response to this line of argument, while appropriate, is not completely satisfactory. He accepts the way in which the problem is defined, and simply argues that the desired conclusion does not follow. His point is purely formal. From the fact that we have a series of finite length, the later elements of which are determined by the earlier, it does not follow that there must a first *in* the series, because the series could be continuous or compact. His example is of an inverted triangle being dipped into the water, the horizontal lines successively being made across the triangle by the water representing cognition and the apex representing the external object. To claim that if there is going to be knowledge of the external object then there must be a first cognition is to claim that when the triangle is dipped into the water there must be a sectional line made by the surface of the water lower or earlier than which no surface line had been made in that way. But it is clear that since any line must be some distance above the apex, it is not true that there must be a first, because there simply are infinitely many. Given the analogy, for any cognition in a series of cognitions of an external object, there would always be a prior one to determine it.[2]

The phrase "given the analogy" is crucial. This counterargument is telling only if we have reason to believe that the series of cognitions is continuous or at least compact. That it is such Peirce does not here argue, but we are given a clue as to his position in his remarks that "cognition arises by a *process* of beginning, as any other change comes to pass" (5.263) and, even more explicitly, "our experience of any object is developed by a process continuous from the very first, of change of the cognition and increase in the liveliness of consciousness" (*W*, 2.191). The point seems to be that coming to know is a genuine process, processes are continuous, and continuous series need not have first members.

He goes on to argue that the foundationalist's positing of such a first member is not only unnecessary but both unintelligible in itself and unhelpful for the understanding of the series. On the first point, Peirce argues that an hypothesis is justified only to the degree that it is able to explain the facts. And, on his view of explanation, to suppose a given cognition (e.g., the first in the series) to be determined by something absolutely external would not qualify as an informative explanation of the determination of that cognition:

> To adduce the cognition by which a given cognition has been determined is to explain the determinations of that cognition. And it is the only way of explaining them. For something entirely out of consciousness which may

> be supposed to determine it can, as such, only be known and only adduced in the determinate cognition in question. So that to suppose that a cognition is determined solely by something absolutely external, is to suppose its determinations incapable of explanation. (5.260)

Thus, such supposition would be a contentless hypothesis. Secondly, even if one could make sense out of the positing of such a first cognition, it would be unhelpful; for the relevant consideration for understanding a series of cognitions is not its first member but the rule by which the series is generated. Such rules are provided by the rules for valid inference.

These are the various strands in what Peirce has to say by way of disengaging the final transcendental argument for intuitions, for foundationalism. I claim that this line of response, while appropriate, seems unsatisfactory for two reasons: First, it insists on posing the justification issue in terms of temporal priority rather than mere logical priority; and secondly, it seems to accept a linear picture of dependence of one cognition on another when a more diffuse network interdependence picture would seem to be more appropriate.

It does appear that Peirce is concerned to argue that there is not only no logically first cognition in any series but no temporally first cognition either. The point seems to be that although the process of coming to know a given object does take place in a finite time, nonetheless given that there is no minimum finite interval for a cognition, there then can be "an infinite series of inductions and hypotheses" (5.311) between any culminating cognition and its external object. Moreover, the picture that is projected by this way of looking at the issue is that of a cognition always being determined by prior cognition in a straight line back to the object. But while Peirce's argument does formally block the specific conceptual argument for foundationalism, this does not seem to be the best way to look at the issue of justification if one's aim is to provide a plausible alternative to foundationalism. A more promising alternative would appear to involve eschewing the linear picture entirely and thinking of the justification of any given cognition in terms of its role in an interrelated network of cognitions, none of which faces the tribunal of experience as an individual but only globally as part of a general cognitive structure. This would seem to be not only the most promising way of confronting the third thesis of foundationalism, i.e., that without some instances of cognitive intuition skepticism would ensue, but also the way most in sympathy with Peirce's own social and historical conception of knowledge.

c. *The Critique of Internalism*

Given the obvious candidates for foundational cognition in the Cartesian tradition, it is not surprising that Peirce's critique of foundationalism naturally involves a more general critique of internalism. The Cartesian internalist maintains that the only real justifiers of belief are cognitive items to which the subject has direct access, and that the only items that meet this condition are the mental states of that subject. The point of entry into Peirce's critique of internalism is obviously his critique of introspection explored above.

On his view we don't have direct access to our internal mental world, and what grasp we do have of this realm is derived from and dependent upon our cognitive grasp of publicly available objects. Our cognitive gaze is first directed outward and only by reflection turned inward, so that our knowledge of ourselves, our mental acts, and the data of consciousness is not immediate but logically parasitic on our more ordinary awareness of the objects around us.

The fundamental line of argument for this broadly externalist perspective on the mental is in the first instance methodological. The considerations are general and basically those of parsimony. Between equally comprehensive alternative accounts of a given range of phenomena we should always prefer that explanation which posits as explanans the fewest kinds of entities (faculties, processes, etc.) and which posits as explanans entities whose properties are continuous with those posited by successful explanatory accounts in related domains.

The general attitude is one of methodological conservatism. We should be expansive rather than revolutionary in that we should extend existing explanatory accounts as far as possible before introducing any radically new accounts, and should prefer those whose posits are closest to (most easily related to) the posits in the entrenched accounts in the region.

Now when this general methodological orientation is focused on the issues bearing on epistemology, Peirce's reflections take the following turn. We are trying to get clear about knowledge and to determine which if any of our beliefs qualify as such. For Peirce it is our claims about manifest properties of middle-sized physical objects in the area and theoretical extensions of these claims that are the prima facie candidates for real knowledge. It is in our effort to understand these that we attribute certain abilities to and processes in knowers. Our knowledge of our own mental states is secondary to and in fact is conceptually parasitic upon our grasp of objects in the external world. Peirce even talks of this knowledge of the mental as a 'inference' from our knowledge of the public:

> We must put aside all prejudices derived from a philosophy which bases our knowledge of the external world on our self-consciousness. We can admit no statement concerning what passes within us except as an hypothesis necessary to explain what takes place in what we commonly call the external world. Moreover, when we have upon such grounds assumed one faculty or mode of action of the mind, we cannot, of course, adopt any other hypothesis for the purpose of explaining any fact which can be explained by our first supposition, but must carry the latter as far as it will go. In other words, we must as far as we can do without additional hypotheses, reduce all kinds of mental actions to one general type. (5.266)

Hence, far from being epistemically privileged, our knowledge of our own mental acts is epistemically secondary. In its paradigmatic instances knowledge is not understood to be private but rather public, so justification is construed as a social process which depends throughout on publicly accessible objects and the possibility of social reinforcement and correction. This point leads naturally into Peirce's final critique of individualism.

d. The Critique of Individualism

The source of Peirce's most fundamental dissatisfaction with Cartesianism is his conviction that its basic orientation is inimical to the social impulse of man and the manifestation of this impulse in the practice of science. Cartesianism locates certainty "in the individual consciousness" and makes "single individuals absolute judges of truth" (5.265). Accordingly, the notions of consensus and disagreement play no epistemic roles, given that justification and confirmation are both private and individual as opposed to public and social. For Peirce, this individualistic conception of knowledge is totally inappropriate, given the paradigmatic instances thereof, namely, the established claims of science.

Peirce's general recommendation for the consideration of issues bearing on knowledge and justification is straightforward:

> Philosophy ought to imitate the successful sciences in its methods, so far as to proceed only from tangible premisses which can be subjected to careful scrutiny, and to trust rather to the multitude and variety of its arguments than to the conclusiveness of any one. Its reasoning should not form a chain which is no stronger than its weakest link, but a cable whose fibers may be ever so slender, provided they are sufficiently numerous and intimately connected. (5.265)

The specific maxims contained in this general recommendation to take science as our model of knowledge are three in number. First, give up the simple linear model of justification, replacing the 'simple thread of inference' picture with a more holistic network model of justification in which there are multiple interrelated lines of consideration bearing on the various elements of the cognitive structure. Secondly, by "proceed only from tangible premisses" he means to suggest that the evaluation of evidence be conducted in the public domain, invoking criteria that are communal bearing on data that are intersubjectively available. Thirdly, we should regard the notions of consensus and disagreement as epistemically decisive.

This third point is really the linchpin in Peirce's attack on Cartesian individualism: "If disciplined and candid minds carefully examine a theory and refuse to accept it, this ought to create doubts in the mind of the author of the theory himself" (5.265). The claim is that this lack of agreement "ought to create doubts." Whether it be individual beliefs or overarching theories, they should be regarded as "on probation until this agreement is reached," with justified conviction and ultimately truth being a matter of achieved consensus. This theme resonates throughout the writings of Peirce and finds its place as one of his maxims of scientific reasoning: "what is questioned by instructed persons is not certain" (*W*, 2.356); "what men cannot come to agreement upon no one of them can be said to know" (*W*, 3.38); "no sensible man will be void of doubt as long as persons as competent to judge as himself differ from him" (*W*, 2.355). This epistemic employment of consensus as a criterion does not grow out of any democratic concerns but rather out of a general conviction about the social nature of inquiry given the finitude of human perspective. Far from democratic, these maxims are firmly tied to the notion of expertise ("as competent to judge") and the fragmentary character of individual understanding. Peirce, of course, would certainly acknowledge that there are some special circumstances where I may be in an epistemically privileged position vis-à-vis some particular fact (e.g., I know that I didn't commit the murder in spite of overwhelming circumstantial evidence to the contrary that should convince others; or lab A discounts the contrary results of lab B upon learning that the latter has failed to carry out what the former knows to be the crucial experiment); but in these cases the nature of the epistemic privilege is uncontroversially understood. Moreover, even in these situations there can be limit cases of appropriate self-doubt, and, most importantly, most of the interesting cases of cognition don't involve this mundane kind of epistemic privilege. Our understanding of how things are (even with regard to ourselves) is not an individual but a communal project.

This critique of individualism points in the direction of science, and in so pointing recapitulates strands in the critiques of methodism, foundationalism, and internalism. The basic contention is that the overall view of knowledge and justification that he calls 'Cartesian' is simply out of step with developments of scientific understanding and explanation. Moreover, he does not see this as a narrow critique: "In some or all these respects most modern philosophers have been, in effect, Cartesians.... it seems to me that modern science and modern logic require us to stand on a very different platform from this" (5.264). Against the backdrop of this multi-faceted critique, I will now examine Peirce's positive epistemological stance.

3. Fallibilism, Certainty, and Critical Common-Sensism

Not surprisingly, the notions of doubt and certainty are on center stage in Peirce's critique of Cartesianism. 'Cartesian doubt' and 'Cartesian certainty' are under serious attack, but it is also important to appreciate the constructive roles the general notions of doubt and certainty play in Peirce's own conception of knowledge. They are focal elements of his fallibilism and his critical common-sensism.

a. Fallibilism, Skepticism, and Certainty

One of the most important keys to an appreciation of Peirce's overall attitude toward human knowledge is his general fallibilistic attitude. In a relatively late manuscript he gives one of his more sweeping statements of fallibilism:

> All positive reasoning is of the nature of judging the proportion of something in a whole collection by the proportion found in the sample. Accordingly, there are three things we can never hope to attain by reasoning, namely, absolute certainty, absolute exactitude, absolute universality. We cannot be absolutely certain that our conclusions are even approximately true; for the sample may be utterly unlike the unsampled part of the collection. We cannot pretend to be even probably exact; because the sample consists of but a finite number of instances and only admits special values of the proportion sought. Finally, even if we could ascertain with absolute certainty and exactness that the ratio of sinful men to all men was as 1 to 1; still among the infinite generations of men there would be room for any finite number of sinless men without violating the proportion.... Now if exactitude, certitude and universality are not to be attained by reasoning, there is certainly no other means by which they can be reached. (1.141–42)

Several things are obvious from this general statement. First, all knowing is deemed to have the same general structure as scientific theorizing. Even the most simple cases of belief formation have the structure of hypotheses whose meaning is to be unpacked in terms of predictions about future experiences. Developed science is simply knowing 'writ large'. Secondly, once we appreciate this fact, fallibilism follows directly from the general logic of confirmation we have previously explored. Beliefs have the nature of entrenched hypotheses that have not yet been falsified or called into question by their clash with experience. As such they command our ever-increasing confidence, but the difference between even the highest degree of warranted confidence and absolute certainty can never be traversed. Since the number of potentially confirming or falsifying experiences relevant to any belief is infinite, we can never rule out the possibility that the belief will have to be eventually discarded.

The intended scope of Peirce's fallibilism can be inferred from the range of objections to it he considers. As ways of knowing claiming to transcend fallibilistic strictures, he singles out knowledge by revelation, the a priori grasp of necessary truths, and the deliverances of direct perception. His rejection of these three as in any way exceptions or counterexamples indicates the intended scope of his fallibilism.

Peirce takes the claims of revelation seriously not only for historical-religious reasons but because testimony as a source of knowledge seems central to all legitimate claims to know. Even granted this positive disposition, however, he does think that "revealed truths ... constitute by far the most uncertain class of truths there are" (1.143). In any event, he maintains that they don't constitute a serious obstacle to fallibilism: First, although many purport to be universal in scope, they are never universally accessible, and this for no apparent internal reason; secondly, most revelation does not even make any pretension to exactitude; finally, although most revelations do claim to warrant certainty, it seems clear that the truth of such a claim can be established only by some form of reasoning and that the non-conclusiveness of this reasoning will infect the degree of assent appropriate even to the original content. Moreover, it seems to be a feature of religious revelation that it have a dimension of incomprehensibility about it (given its source), so that we can never be sure that we rightly comprehend it or that any of our formulations precisely capture it. If for no other reason, humility would seem to require a certain fallibilism toward the content of revelation "even if revelation was much plainer than it is" (1.143).

Peirce next considers the claims on behalf of a priori insight to an infallible grasp of the axioms of geometry, the principles of logic, the maxims of causality, and the like. He first simply points out that the weight of historical evidence seems to count against the view that any particular instance of such a claim is absolutely certain, without exception, and exact. There is certainly no historical consensus on particulars. To those who do not consider consensus or lack thereof as relevant, he asks "how does one know that these a priori truths are certain, exceptionless and exact?" An answer in terms of 'reasoning' would seem to undermine the certainty, and an answer itself in terms of 'a priori insight' would beg the question.

The last alleged exceptions to the structures of fallibilism are the claims of direct perception. As an objection to a general claim about the status of beliefs, i.e., fallibilism, this objection would have to be more precisely formulated in terms of perceptual judgments, and Peirce maintains that there is no in-principle difference between perceptual judgments and scientific hypotheses:

> Abductive inference shades into perceptual judgment without any sharp line of demarcation between them; or, in other words, our first premisses, the perceptual judgments, are to be regarded as the extreme case of abductive inferences from which they differ in being absolutely beyond criticism. The abductive suggestion comes to us like a flash. It is an act of *insight*, although of extremely fallible insight. (5.181)

Perceptual judgments are special but only because they are beyond criticism, not because they are infallible. Given that none of these three ways of knowing stands as a serious counterexample to fallibilism, Peirce concludes the discussion sweepingly: "We can never be absolutely sure of anything, nor can we with any probability ascertain the exact value of any measure or general ratio" (1.147).

While this fallibilistic attitude has its theoretical roots in Peirce's general conception of the structure of knowledge, it has its concrete roots in his experience as a working scientist. A good percentage of his working life was spent as an experimentalist, working in what he felt was one of the most precise corners of one of the most precise sciences. Even here, "no man of self-respect ever now states his result without affixing to it its probable error" (1.9). So impressed was Peirce with the fact that even in the exact sciences of measurement our efforts in principle result only in approximation that he devoted some of his earliest theoretical work to the study of what he called 'the theory of errors'. His conviction was that the fallibilistic attitude was in order even with regard to our most precise efforts in this circumscribed domain; the magnitude and probability of error only increased as one moved to the more theoretical reaches of science. And since science is our paradigmatic instance of knowledge, that most amenable to responsible control, it would seem reasonable to believe that our more speculative or general cognitive endeavors would be attended with an even greater degree of and probability of error. Fallibilism is an attitude of mind that should range over all our cognitive endeavors: "I will not, therefore, admit that we know anything whatever with *absolute certainty*" (7.108).

If fallibilism, then, amounts to a rejection of absolute certainty, it is equally committed to the rejection of skepticism. Certainty and skepticism are nurtured by the same assumption, namely, that there is a non-perspectival Archimedean point on which one can ground everything or from which one can question everything. On Peirce's account, on the contrary, our only cognitive situation is *in medias res*, whence some beliefs are called into question on the basis of others which are taken for granted. Any functional notion of an 'absolute perspective' is eschewed; its surrogate, that of the final community, being a limit notion cannot play this active cognitive role.

Modern science does not take general skepticism seriously; nor did Peirce. Far from its being the case that we can doubt everything, or that it might be the case that we know nothing, Peirce maintained that there were in fact quite a few things that we did know about the world:

> Upon innumerable questions we have already reached the final opinion. How do we know that? Do we fancy ourselves infallible? Not at all; but throwing off as probably erroneous a thousandth or even a hundredth of all the beliefs established beyond present doubt, there must

> remain a vast multitude in which the final opinion has been reached. Every directory, guide-book, dictionary, history, and work of science is crammed with such facts. (8.43)

Although we are never in a position to justify the claim that we have the absolute truth on any given matter, in point of fact the appropriate consensus may have been achieved and we may in fact have attained the truth. Many of the very general (and vague)beliefs we have about ourselves and our relations to our proximate environment may well fall into that category, as may some well-confirmed facts of history and the regularities and laws of the basic sciences. Having attained the truth in many cases is not incompatible with fallibilism.

Even further, Peirce is not only willing to allow but insists that we do entertain certain propositions that are beyond question and others that are infallible. First, there is a sense in which normal perceptual judgments are beyond question, since they are brought about at a level beyond our critical control. It, of course, does not follow from this sense of 'beyond question' that they are true. Secondly, again displaying an attitude the very antithesis of skepticism, Peirce is willing to say of scientific inquiry itself that it is infallible, inasmuch as it "is foredestined to reach the truth of every problem with as unerring an infallibility as the instincts of animals do their work" (7.77) and "has never faltered in its confidence that it would ultimately find out the truth concerning any question in which it could apply the check of experiment" (7.87). These, however, are not claims about the world but quasi-formal claims about the nature of scientific inquiry. Nonetheless, if skepticism is an attitude of cognitive despair, Peirce's fallibilism seems to be perfectly compatible with robust cognitive hope.

b. Critical Common-Sensism

The positive epistemological stance that emerges from this critique of Cartesianism Peirce calls "critical common-sensism." His choice of terminology derives from the fact that he does think that our cognitive interaction with the world is grounded in a range of common sense beliefs that are beyond question. These common sense beliefs, however, are not contrary to science but rather continuous with that mode of cognition which is developed science. In fact, his general view about the structure of all knowledge is modeled on his view about the structure of the developed sciences. The general notions of belief, evidence, justification, and truth are explicated in terms of criteria that are public and social, rather than private and individual, because the aim of cognition is understood to be objective knowledge not subjective certainty. Furthermore, the activity of knowing is seen as intrinsically communal, and the cognitive product of the activity is understood to be a developing historical reality structured to withstand continual revision over time all the way down. The general picture of knowledge that emerges is very different from the picture that motivated Descartes.

The view that our overall system of beliefs about the world is grounded in a range of common sense beliefs that are beyond question calls for some explication. "Beyond question," as we have seen, does not imply truth nor does it mean to suggest that we will never have a reason to call specific instances of this class of beliefs into question. It means simply that there are *particular* beliefs that we have in specific concrete circumstances that are brought about in us below the level of critical reflection and, secondly, that there are *kinds* of beliefs we have that it would be hard to imagine having a reason globally to reject, because they are caused by the common insistency of everyday experience over many generations of multitudinous populations.

Peirce's attitude toward this set of indubitable perceptual beliefs is quite nuanced.[3] Strictly speaking these basic perceptual beliefs can and do change over time and from individual to individual. The changes, however, are "so slight from generation to generation" (5.444) that it seems not inappropriate to think of a fixed list of such basic beliefs which would be for all practical purposes the same for all people. If we were to think of these perceptual beliefs in this 'fixed list' way, however, we should not suppose "it is absolutely fixed ... but that it is so nearly so that for ordinary purposes it may be taken as quite so" (5.509).

That there should be such an uncritical consensus at the basic perceptual level is understandable in terms of both the function of perceptual beliefs in organic survival and the status of these indubitable perceptual beliefs as invariably vague. It would seem to be an evolutionary necessity that ordinary perceptual judgments made in standard conditions be reliable guides to and through the practically relevant features of the world around us. It is a post-facto necessity in the sense that our specific survival can be explained in terms of the relative reliability of our perceptually guided interactive mechanisms. Moreover, the vagueness of the perceptual beliefs contributed to their practical reliability. Given their evolutionary role, they are geared to the level of discrimination necessary for survival in a primitive mode of life. At this level the trade-off between speed (both of habit-acquisition and triggered response) and fine-grained accuracy is heavily in favor of the former. The cost in terms of danger of any improvement in accuracy has to be figured into the equation.

Hence the indubitability and reliability of our basic perceptual judgments, a principal rationale for the designation 'common-sensism', is circumscribed by this understanding of their scope and function. Peirce insists that his common-sense philosophical orientation is to be distinguished from the Scottish variety by the adjective "critical," which serves the double purpose of indicating both a positive attitude toward critical revision, even at the common-sense level, and the continuity of these common-sense convictions with the more fine-grained theoretical conclusions of developed science. He identifies as 'clauses' of critical common-sensism the following:

> That there are indubitable beliefs which vary a little and but a little under varying circumstances and in distant ages; that they partake of the nature of instincts, this

> word being taken in the broad sense; that they concern matters within the purview of the primitive man; that they are very vague indeed (such as, that fire burns) without being perfectly so; that while it may be disastrous to science for those who pursue it to think they doubt what they really believe, and still more so really to doubt what they ought to believe, yet, on the whole, neither of these is so unfavorable to science as for men of science to believe what they ought to doubt, nor even for them to think they believe what they really doubt; that a philosopher ought not to regard an important proposition as indubitable without a systematic and arduous endeavor to attain to a doubt of it, remembering that genuine doubt cannot be created by a mere effort of will but must be compassed through experience; that while it is possible that propositions that really are indubitable, for the time being, should nevertheless be false, yet insofar as we do not doubt a proposition we cannot but regard it as perfectly true and perfectly certain; that while holding certain propositions to be each individually perfectly certain, we may and ought to think it likely that some one of them, if not more, is false. (5.498)

This passage alludes to many facets of the general epistemological picture Peirce has in mind. First, there is a relatively fixed list of indubitable perceptual beliefs of the nature of instincts that guide our fundamental interactions with our environment. Secondly, these are essentially vague but determinate enough for practical purposes, and hence their primary epistemic authority is limited to those levels of gross interaction with the world that are relevant to our practical purposes. Thirdly, as our purposes become more theoretical and refined we move beyond the sphere of their principal authority, so that critical scrutiny of them becomes both possible and appropriate. Fourthly, even with regard to our highest theoretical activities, these basic perceptual beliefs play the indispensable roles of fixing the given subjects of scientific inquiry and identifying the ranges of phenomena that can confirm or falsify our various scientific speculations. The relationship between developed scientific understanding of the world and our primitive beliefs about it is not that of the true to the false but of the precise to the vague. Far from there being an antithetical relationship between our primitive conception of the world and our scientific conception, there is a symbiosis because scientific theorizing takes place within our ordinary perceptually guided interactions with the world in such a way that the epistemic status of both are inextricably intertwined. Fifthly, real serious doubt even of what we have identified as basic perceptual beliefs should be seen to play an indispensable role in cognitively responsible inquiry. It is not only possible that what we correctly regard as indubitable is false, but we ought to think it likely that some of our indubitable beliefs are in fact false.

The Cartesian demand for an absolute foundation for knowledge has been left far behind, and new metaphors for our overall system of belief naturally suggest themselves. Whereas Descartes' image was both linear and static, Peirce's is both holistic and developmental. These two features suggest other metaphors for understanding and exhibiting epistemic structure. Three that come readily to mind are Neurath's ship, Popper's building in a swamp, and Quine's fabric or force field.

Neurath compares our relation to our system of beliefs to that of "sailors who must rebuild their ship in the open sea, never able to dismantle it in dry-dock and to reconstruct it there out of the best materials."[4] Any part can develop problems while at sea, and we have no place to stand while making repairs other than on the very ship being repaired. Given the corruptible nature of the material the process is without end; and, inasmuch as we were born on the ship, for all we know it was without beginning. Some parts of the ship are obviously more difficult to repair than others given their pivotal structural roles, but they too may call for repair. The point of the metaphor is that we have to correct and develop our system of belief from within. We begin our reasoning within an inherited system of beliefs and whatever corrections and additions we make are justified from within that system. There is no dry-dock; nor can we jettison the ship and start anew. It doesn't follow from this systemic or holistic view of belief that there are no differences in kind among our beliefs; in fact, what Neurath calls protocols, i.e., those reports having to do with our direct experience of objects, do play a very special role. But even they are subject to verification and are revisable. Moreover, when we view the system as a whole there is a difference between mere tinkering and radical revision, but in both cases we are talking about degrees of internally based repair.

Popper's metaphor generalized from science to our whole system of belief is equally suggestive:

> The empirical basis of objective science has nothing 'absolute' about it. Science does not rest upon rock-bottom. The bold structure of theories rises, as it were, above a swamp, but not down to any natural or 'given' base; and when we cease our attempts to drive our piles into a deeper layer, it is not because we have reached firm ground. We simply stop when we are satisfied that they are firm enough to carry the structure, at least for the time being.[5]

For our purposes there is no bedrock; it is swamp all the way down. If, however, we are knowledgeable engineers we can construct a system of piling sufficient to give stability to our upper-story construction, and this in turn, if adequately balanced, can add additional stability to the piling. Our overall system of belief may well have a structure such that it is meaningful to distinguish between the analogues of the building proper and the piles, but there is nothing absolutely sharp about the distinction and no rock bottom to the structure.

Finally, there are Quine's metaphors of the fabric of belief or the field of force:

> The totality of our so-called knowledge or beliefs, from the most casual matters of geography and history to the

> profoundest laws of atomic physics or even pure mathematics and logic, is a man-made fabric which impinges on experience only along the edges. Or, to change the figure, total science is like a field of force whose boundary conditions are experience. A conflict with experience at the periphery occasions readjustment in the interior of the field. Truth values have to be redistributed over some of our statements.... But the total field is so underdetermined by its boundary conditions, experience, that there is much latitude of choice as to what statements to re-evaluate in the light of any single contrary experience. No particular experiences are linked to any particular statements in the interior of the field, except indirectly through considerations of equilibrium affecting the field as a whole.[6]

In this picture of the structure of knowledge our various kinds of beliefs form an interrelated network which faces the tribunal of experience as a whole rather than atomistically. The network of beliefs has to be readjusted continually in the face of recalcitrant experience, but there is no unique way in which the adjustment has to be made. The clash is at the level of experience, and observation sentences are on the front lines; but the accommodation of the system can take different forms and the adjustment may well be made at a different level. No belief is in principle unrevisable, and any belief can be held firm if one is willing to pay the price elsewhere in the network. Quine, in contrast to Peirce, thinks that these factors militate against convergence on a unique set of beliefs at the limit of inquiry.

All three metaphors contrast sharply with the Cartesian image of knowledge as an edifice built on an absolute unshakable foundation, but they are not meant to suggest a pure coherentist picture either. Not all kinds of beliefs play the same roles, nor are all beliefs equally solid or equally vulnerable. There are the beams of the ship, the piles in the swamp, and the periphery of the fabric or the force field. Neurath's protocols, Popper's observations, and Quine's observation sentences, while certainly not indubitable or unrevisable, do play special roles in the cognitive enterprise and do have a level of prima facie credibility that enables them to play these roles.

Peirce's critique of Cartesianism stands as a classic predecessor to this later tradition, and his positive picture of our overall cognitive structure is similarly nuanced. The image is at the same time holistic and so structured as to be grounded in perception. It is the insistency of perceptual experience that prompts and guides continual revision, yet the epistemic status of the reports of perceptual experience is not independent of the overall system of belief:

> Our perceptual judgments are the first premisses of all our reasonings.... All our other judgments are so many theories whose only justification is that they have been and will be bourne out by perceptual judgments. (5.116)

This structure licenses the categorization of Peirce's view as an "empiricism," but it is a post-Kantian empiricism, an empiricism without classical foundations. Obviously, the plausibility of such a view of the structure of knowledge rests heavily on its account of perception. Peirce was not blind to this fact nor was he negligent in the execution. He saw that his view of the structure of science in particular and of cognition in general called for a detailed account of the nature of perceptual judgment that would render intelligible the distinctive role of the insistency of experience in the development of knowledge. His general theory of meaning and his intersubjective view of evidence, confirmation and truth ultimately depend on it. On the assumption that pure coherentism is not a viable alternative to Cartesian foundationalism, the plausibility of his critique of Cartesian foundationalism is tied to his positive account of the epistemic role of perception.

c. *The Epistemic Role of Perception*

In exploring in detail the status of perceptual judgments in the cognitive process, we need first to examine the phenomenological features of perceptual experience and also provide some analysis of the structural elements of the corresponding judgments that enable perceptual judgments to play such a primary role in knowledge acquisition and validation. Secondly, against the background of an understanding of the nature and primacy of perceptual judgments, the question of their revisability has to be explored. Finally, given the primary role of such judgments in keeping a given inquiry on track, we need some account of what it is about the semantic structure of perceptual judgments that enables them to play such a monitoring function vis-à-vis inquiry. Fortunately Peirce explores these three facets of the issue in some detail, starting with the most general features of perceptual experience.

Peirce's principal criticism of the idealist tradition is that it does not appropriately acknowledge the phenomenological fact that perceptual experience presents itself as the experience of a world that is independent of the perceiver and with which the perceiver is confronted. Peirce calls our attention to the dyadic character of perceptual experience:

> It involves the sense of action and reaction, resistance, externality, otherness, pair-edness. It is the sense that something has hit me or that I am hitting something; it might be called the sense of collision or clash... The capital error of Hegel which permeates his whole system in every part of it is that it almost altogether ignores the Outward Clash. (8.41)

Perceptual experience includes as one of its features the otherness of the objects perceived. To perceive is to encounter a world not of our own making, a world forced upon us to which we must make an accommodation. Peirce felt that both the classical empiricists and idealists downplayed this dyadic 'compulsive' dimension of perceptual experience in their over-emphasis on the features of felt immediacy or meaning respectively. No matter how immersed in either feeling or theory, "we are continually bumping up against hard fact; we expected one thing or passively took it for granted and had the image of it in our minds, but experience forces that idea into the background and compels us to think quite differently" (1.324).

The notion of perceptual experience is a global notion that not only admits of but calls for a detailed analysis into its elements if we are going to get any satisfactory answers to the epistemological questions with which we are concerned.[7] It is theoretically decomposable into simpler elements, but the analysis should not blind us to the holistic character of the experience itself. Although it is not inappropriate to talk of *this* particular perceptual experience and these *components* of this particular experience, our actual perceptual experience is not a series of discrete units made up of isolatable parts but rather a continuous flow. The actual experience, no matter how direct or how short, involves dimensions of confrontation and meaning as well as elements of memory and anticipation. However, this having been said, Peirce acknowledges the legitimacy of analysis and the significance of abstractly characterizing the various structural elements of perceptual experience.

In fact, he developed a methodology for considering in isolation features of phenomena which are not in fact separable. He called the operation 'precision' and described it as an act of mental abstraction which "arises from *attention to* one element and *neglect of* the other" (1.549). When this analytic intention is focused on the flow of perceptual experience, Peirce is able to distinguish as elements the percept, the percipuum, and the individual perceptual judgment. The paragraph in which all three notions are introduced calls for careful scrutiny:

> We know nothing about the percept otherwise than by testimony of the perceptual judgment, excepting that we feel the blow of it, the reaction of it against us, and we see the contents of it arranged into an object in its totality.... But the moment we fix our minds upon it and *think* the least thing about the percept, it is the perceptual judgment that tells us what we so "perceive." For

> this and other reasons, I propose to consider the percept as it is immediately interpreted in the perceptual judgment, under the name of the "percipuum." (7.643)

As one prescinds the elements from the concrete flow of perceptual experience, the order is from the *perceptual judgment*, through the *percipuum*, to the *percept* as one moves away from the complex phenomenon of meaningful perceptual experience toward what simply confronts one in perception. The movement, however, is one of abstractive analysis, not an attempt to identify anything cognitively given.

The percept is the most primitive sensory presentation of the object characterized by the relevant sensory qualities and by the feeling of opposition or otherness. In itself it is a determinate particular. Peirce is willing to characterize it as an *image* which "obtrudes itself upon me in its entirety," as long as we do not build into the notion of an image the role of representation:

> [The percept] makes no professions of any kind, essentially embodies no intentions of any kind, does not stand for anything. It obtrudes itself upon my gaze; but not as a deputy for anything else, not "as" anything. It simply knocks at the portal of my soul and stands there in the doorway. (7.619)

The percept, then, is the bare compulsive experience composed of qualities of feeling independent of relations and interpretation. It is something that happens to me, not something I construct, although it undoubtedly is itself something constructed out of more primitive informational input below the level of consciousness: "every percept is the product of a mental process ... and these processes are of no little complexity" (7.624). We, of course, are not aware of this process, and its structure is a matter of psychological speculation. Peirce sometimes refers to this process as that bearing on the generation of "the first impressions of sense" (2.141). Furthermore, percepts are neither believed nor disbelieved, certain nor uncertain, true nor false; they are simply directly experienced. They are forms of sensory awareness which prompt, and in turn are described by, perceptual judgments.

Hence, while percepts are our most primitive conscious informational input, it is perceptual judgments that are our most primitive *cognitive* units. A perceptual judgment is initially defined as "a judgment asserting in propositional form what a character of a percept directly present to the mind is" (5.54). It is the act of forming a mental proposition about some characteristic of the perceptually given, together with an assent to this proposition. Peirce notes several important differences between the percept and the perceptual judgment. First, taking as an example the perception of a yellow chair, the percept presents the chair as a unity involving no analytic elements whatever, whereas the perceptual judgment separates the chair from the color, making one the predicate of the other (7.631). Secondly, the 'singularity' of the percept is definite with regard to this individual and this shade of yellow while the generality of the predicate in the perceptual judgment allows it vaguely to range over all yellow things and all shades of yellow (7.632–33). Thirdly, the perceptual judgment no more resembles the percept than a description of a painting resembles the painting itself (5.54), but is related to its percept as an index that arises from it dyadically (7.628). Fourthly, a percept is non-propositional in form and thus can stand in no logical relation to anything else; only the perceptual judgment is capable of sustaining logical relations (7.628). Hence, although the percept might have a more primitive claim to the title 'evidence' of the senses, it is only the perceptual facts—"the intellect's fallible record of the percepts"—which function cognitively in reasoning both as starting points and as controls.

To this point Peirce's account of perception seems straightforward, but the details of the account from this point on are considerably less clear, both for intrinsic and developmental reasons. The issues themselves are very complex and to deal with them he introduces an additional complication; the original analysis was merely in terms of the percept and the perceptual judgment, while later discussions include a third category, the 'percipuum', in the analysis. However, there is a discernible line of thought which is Peircean and may even be his.

The stark differences between the percept and the perceptual judgment pose a problem: how is the former 'taken up into' the latter; how does the percept function in the perceptual judgment? An element of Peirce's response is to introduce an intermediary from which our concept of the pure percept is really an abstraction. The basic unit of which we are cognitively aware is not the percept itself, but the percept as immediately interpreted in the perceptual judgment, for which he proposes the term "percipuum":

> We know nothing about the percept otherwise than by testimony of the perceptual judgment, except that we feel the blow of it, the reaction of it against us, and we see the contents of it arranged into an object, in its totality,—excepting also, of course, what the psychologists are able to make out inferentially. But the moment we fix our minds upon it and *think* the least thing about the percept, it is the perceptual judgment that tells us what we so "perceive." For this and other reasons, I propose to consider the percept as it is immediately interpreted in the perceptual judgment, under the name of the "percipuum." The percipuum, then, is what forces itself upon your acknowledgement, without any why or wherefore, so that if anybody asks you why you should regard it as appearing so and so, all you can say is, 'I can't help it; that is how I see it'. (7.643)

We come to know facts about our world by means of the perceptual judgment which, through the percipuum, indicates the percept which indicates the physical object.

The percipuum, then, is the percept as immediately interpreted, and the operative notion here is *immediately*. It does not follow from this that the only thing we *directly* know is the percipuum or even the percept, or that we only know indirectly via inference the characteristics of independent physical objects. In fact, Peirce uses as examples of direct perceptual judgments both "this chair appears yellow" (7.631) and "this chair is yellow" (7.635)—characterizations of things as they appear to us and characterizations of things *simpliciter*.

Under ordinary circumstances the paradigmatic perceptual judgment is of the second kind: it is a direct report about the sensory characteristics of things *which* present themselves in our perceptual field but not merely *as* they so present themselves. However, our judgments about them, while *direct* (i.e., not an inference from judgments about percepts) is not unmediated. The ground of our full-blown perceptual judgment is a more primitive judgment about how things appear to us, but we don't infer the former from the latter. In fact, in most cases we infer the latter from the former, i.e., we reason that it must be on the basis of some normally unattended-to primitive recognition of how things appear to me that I am prompted to make the judgment that, e.g., the chair is yellow. This is surely not a conscious inference, nor is there any entailment involved, as is made clear by cases of hallucination. In cases of hallucination the percipuum is exactly the same as in cases of veridical perception, it is just that the corresponding higher order perceptual judgments are called into question by their failure to cohere with other relevant perceptual judgments, both of ourselves and others. Hence, the chair may well appear yellow to us even in cases where it is not, but in those cases where we correctly perceptually judge that the chair is yellow it is on the basis of its appearing yellow to us.

Peirce's account obviously involves a liberal employment of the notion of *unconscious inference*. In fact, Peirce is quite specific about the precise logical form of perceptual judgments, namely, they are to be regarded as limit cases of abductions:

> The perceptive judgment is the result of a process ... [and] if we were to subject this subconscious process to logical analysis, we should find that it terminated in

> what that analysis would represent as an abductive inference. (5.181)

Perceptual judgments are to be thought of on the model of the ascription of a general predicate to individuals, which would reduce them to some kind of unity and thereby render them intelligible. They have the form of hypothetical interpretations of given elements and are general in nature.

It is important to note, however, that when we are speaking of perceptual judgments as abductions we are speaking analogously, because these instances of abductions are both subconscious and uncontrolled, characteristics contrary to standard abductions. Strictly speaking, perceptual judgments are not really judgments that we make but rather ones that are forced upon us:

> Even after the percept is formed there is an operation which seems to me quite uncontrollable. It is judging what it is that a person perceives.... I do not see that is possible to exercise any control over that operation or to subject it to criticism.... Consequently, until I am better advised, I shall consider the perceptual judgment to be utterly beyond control. (5.115)

The kind of perceptual judgment he is talking about here involves a characterization of what he later called the percipuum, and of this he explicitly says it is "a recognition of a character of what is past, a percept which we think we remember; the interpretation is forced upon us but no reason for it can be given" (7.677). He also holds that perceptual judgments more generally speaking are beyond our control, are forced upon us. In the appropriate circumstances we find ourselves judging that the chair is yellow, not merely that it appears yellow to us. In the relatively momentary concrete existential circumstance, even this complex perceptual judgment is beyond our critical control.

It is also important to be clear about the reason why these judgments are beyond our control. They are not logically different from other judgments we make; as do other judgments, they involve leading principles which guide acts of classification. It is just that the relevant mental operations are triggered automatically and subconsciously, and thus are not subject to critical control:

> You may adopt any theory that seems to you acceptable as to the psychological operations by which perceptual judgments are formed.... All that I insist upon is that these operations, whatever they may be, are utterly beyond our control and will go on whether we are pleased with them or not. (5.55)

In the appropriate concrete circumstances these perceptual judgments are things that happen to us, not things we do.

In virtue of this compulsion these judgments are indubitable: "If one *sees* one cannot avoid the percept; and if one *looks* one cannot avoid the perceptual judgment" (7.627). In the presence of the percept we are impotent to refuse our assent to the corresponding perceptual judgment: "The interpretation is forced upon us but no reason for it can be given" (7.677). In concrete circumstances perceptual judgments are indubitable, and their status as indubitable allows them to play a special role in inquiry:

> It follows, then, that our perceptual judgments are the first premisses of all our reasonings and that they cannot be called into question. All our other judgments are so many theories whose only justification is that they have been and will be borne out by perceptual judgments. (5.116)

Although logically speaking they have the structure of hypotheses themselves, perceptual judgments enter the inquiry process as 'first premises' providing both the suggestion for and the test of other levels of inference. Peirce offers as his interpretation of the classical principle *Nihil est in intellectu quod non pruis fuerit in sensu* the stipulation that "the perceptual judgment is the starting point or first premiss of all critical and controlled thinking" (5.181). As de facto indubitable perceptual judgments provide both a springboard and a touchstone for the more speculative and critical dimensions of thought.

This compulsion and primacy, however, do not add up to ultimate epistemic authority. On Peirce's account perceptual judgments are both fallible and revisable. About the fallibility Peirce is quite explicit: "We all know, only too well, how terribly insistent perception may be; and yet for all that, in its most insistent degrees, it may be utterly false" (7.647), and even of the percipuum he says "there is no percipuum so absolute as not to be subject to possible error" (7.676).

The senses in which they are revisable are more subtle. Those perceptual judgments which delineate sensory characteristics of independent physical objects are revisable in the sense that their entailments may conflict with more broadly attested to subsequent perceptual judgments, leading to the conclusion that the earlier ones were hallucinatory. But even the percipuum, which is involved in these perceptual judgments, if not strictly speaking revisable, can in some special cases by controlled. Peirce uses as an example the 'Schroeder's Stair' perceptual illusion. When you first look at the representation you seem to be looking at a flight of stairs from above and cannot see it otherwise. After a few minutes of continual gazing the back wall jumps forward and now you seem to be looking at the stairs from below and cannot see them otherwise. However, if we continually repeat the experiment we can eventually get control over the phenomenon and determine in a particular case how things will appear to us. Thus, we will have "converted an uncontrollable percipuum into a controllable imagination by a brief process of education" (7.647). This case shows what can be done in principle, but is not an indication of how perceptual information is normally processed by the human knower.

Given the crucial role played by perceptual judgments in Peirce's account of science in particular and of knowledge in general, a final word should be added about his account of their semantic structure. While all judgments have an element of generality and thus necessarily employ what Peirce calls tokens (general terms), perceptual judgments are singular in nature and as such involve reference to individuals which can only be indexical:

> Tokens alone do not state what is the subject of discourse; and this can, in fact, not be described in general terms; it can only be indicated. The actual world cannot be distinguished from a world of imagination by any description. Hence, the need of pronouns and indices, and the more complicated the subject the greater the need of them. (3.363)

Since perceptual judgments, then, purport to be about objects in the real world, reference is secured by the indexical component of the judgment, proximately to the percept and ultimately to the independent physical object. Hence Peirce's concern with demonstratives.

The indexical component of the perceptual judgment (proper name, pronoun, demonstrative) fixes the object of the original judgment, and the entity so indicated can then be the object of successive interpretations. The 'immediate' object is the thing as directly experienced by me, but the successive interpretations of this very object may involve the ascription to it of properties quite removed from my immediate experience. This fixing of reference through the indexical element in the judgment is the key to understanding Peirce's view of the history of science as the progressive development of a series of more and more adequate interpretations of what we originally encounter via perceptual judgments.[8]

This understanding of the role of indexicals in perceptual judgment enables Peirce to have a coherent view of empirical knowledge in which perceptual judgments are both primary and revisable, yet the scientific account of the world describes the world as it really is. Our initial interaction with the world is guided by perceptual judgments but subsequent inquiry may give us reason to revise our conception of the objects as given in these judgments. Peirce's distinction between the immediate interpretation of a given object and the final interpretation is saved from sheer discontinuity by the view that it is the indexical component of the judgments that fixes the reference, thus enabling the scientific account to be about the real properties of the very objects originally encountered in perception. Perception can be both the origin and the touchstone of scientific inquiry, yet the final characterization of objects as they really are can be quite different from their initial characterization in perception. This difference, however, is not that between the true and the false or the scientific world and the perceived world (as if there were two different worlds), but rather between the vague characterization of the features of our world sufficient for immediate practical purposes and the more specific and nuanced final characterization of the same world.

IV. *Mind and Reality*

Two very general features have emerged from our exploration of Peirce's thought to this point. First, we have seen that Peirce's general theory of knowledge takes scientific knowing as its paradigm case and develops an account of discovery and justification consonant with both scientific practice and the history thereof. Secondly, the case has been made that his epistemology is broadly speaking in the empiricist tradition, ascribing a primary role to perception both in explicating the meanings of theoretical terms and in the justification of our more theoretical speculations. His is, however, a decidedly post-Kantian empiricism, since there is no level of genuine knowing more basic than judgment, and even the perceptual judgments which are basic are neither self-evident nor incorrigible. Moreover, his developmental holism goes well beyond Kant in insisting on the systematic interrelationship of all our judgments and in acknowledging the changes our belief system is forced to undergo, even in its most basic features, under the pressure of experience over time.

There is, however, more to an account of knowledge than epistemological considerations, so our exploration of Peirce's views on knowledge requires an excursion into more general features of his philosophy of mind. Since the knowing we are concerned with is a mental activity of human beings, the account would be incomplete without a delineation of Peirce's theory of the semiotic structure of mental activity and his more general reflections on the qualitative dimensions of mind. This discussion will culminate in a brief indication of his more speculative metaphysical views on the nature of mind and reality.

1. The Structure and Content of Mental Activity

a. The Structure of Mental Activity

The most crucial commitment in any philosophy of mind has to do with methodology, and Peirce makes this initial move reflectively and systematically. The two methods for exploring our mental life that he considers are, first, introspection, and secondly, theory construction based on external observation. The method of introspection was discussed in detail by Peirce in its relation to his critique of intuition. By "introspection" he meant the direct survey of our own mental life in no way derived from or mediated by our grasp of external, publicly available facts. He argued that we have no such ability and that our direct cognitive awareness was of the characteristics of publicly available middle-sized physical objects, and that it was a mark of 'returning reason' to grasp features of our own mental states (5.247).

His principal argument against the method of introspection is itself methodological in form and is embedded in his more general views about methodology. The principle of parsimony is central to Peirce's views on method: between equally comprehensive alternative accounts of a given range of phenomena we should prefer that explanation which posits as explanans the fewest different kinds of entities (faculties, processes, structures, etc.) and that which posits as explanans entities whose properties are continuous with those already posited by successful explanatory accounts in related domains. His general attitude is one of methodological conservatism: we should be expansive rather than revolutionary, in that we should extend existing explanatory accounts as far as possible before introducing any radically new kinds of explanation. Moreover, in introducing new accounts we should prefer those whose posits are closest to (most easily related to) the posits of the entrenched accounts in the region.

When this general attitude of methodological conservatism is focused on the specific case of the philosophy of mind, his reflections take the following direction. In philosophy of mind we are interested in a comprehensive account of human mental activity including willing, intending, sensing, perceiving, and, principally, thinking. The clearest cases of such mental activity seem to be the public ones, i.e., the thinkings-out-loud, the willings-out-loud and the intendings-out-loud, particularly when these are directed at middle-sized physical objects in the environment. Examples would be: 'that cat is large', 'I want that food', and 'I'll get that spear'. It is with these cases that we should start our account and it is these cases that should be paradigmatic throughout. Moreover, of these public instances of mental activity it seems to be a fact of intellectual history that it is the thinkings-out-loud that are most basic and have gotten the most attention, so it is these that should be our initial point of focus. In summary, then, it is the external rather than the internal 'mental' performances that should be our point of departure and our paradigms; and, even of these, the most perspicuous examples should play the central role, i.e., the thinkings-out-loud.

Now the issue takes the following form. Given that everyone must admit that we have those general cognitive abilities necessary to explain our grasp of public facts about our world, do we have to assume any different cognitive abilities, powers, or faculties to account for our knowledge of our own mental states? The burden of proof is shifted to the one proposing the power of introspection; if our knowledge of our mental states can plausibly be construed as simply a matter of inference from external facts, this explanation is to be preferred. The postulation of the power of introspection would be seen to be unnecessary: "There would be no reason for supposing a power of introspection; and, consequently, the only way of investigating a psychological question is by inference from external facts" (5.249).

When his general methodological considerations and his specific critique of introspection are brought together, Peirce draws the following conclusion:

> We must put aside all prejudices derived from a philosophy which bases our knowledge of the external world

> on our self-consciousness. We can admit no statement concerning what passes within us except as an hypothesis necessary to explain what takes place in what we commonly call the external world. Moreover, when we have upon such grounds assumed one faculty or mode of action of the mind, we cannot, of course, adopt any other hypothesis for the purpose of explaining any fact which can be explained by our first supposition, but must carry the latter as far as it will go. In other words, we must, as far as we can do so without additional hypotheses, reduce all kinds of mental action to one general type. (5.266)

Two methodological constraints dictate the broad contours of Peirce's philosophy of mind. First, since direct introspective access to mental activity is not admitted, internal mental entities and processes are to be posited only as needed to explain the mind's manifest behavior. Secondly, once one kind of mental entity, process, or structure is so introduced, others can be introduced only as needed to explain facts impossible to account for on the first supposition.

Concentrating, then, on thinking-out-loud as the focal mental activity, what are the paradigmatic instances of it, and what features of this paradigm are of principal relevance? Peirce contends that "if we seek the light of external facts, the only cases of thought which we can find are of thought in signs" (5.251). It is, then, our public sign system, i.e., natural language, which should be our point of departure; and it is the structural features of this language which would seem to be of most relevance. Moreover, since thinking is basically a process rather than an achievement, it would be the structural features of this process that would seem to be the key. This process of thought most minimally described would be simply a matter of one thought "following after" another, but Peirce thinks that the deep structure of the process involves one thought "following from" another in patterned ways. This being the case, following his own methodological guidelines, he assumes that the patterns are the patterns of drawing inferences broadly conceived, and that the structure of inference is thus a prime candidate for the structure of thought.

Peirce provisionally draws these conclusions and then extrapolates. Our most characteristic kind of mental activity is thinking and the clearest instances of thinking are instances of thinking-out-loud. As for structure, the most perspicuous of our thinkings-out-loud are the processes of drawing inferences, and the structure of these is logical. Accordingly, it is reasonable to begin our account with the supposition that the principles of logical inference give us the structure of all mental activity and develop this hypothesis as far as we can: "We must as far as we can, without any other supposition than that the mind reasons, reduce all mental action to the formula of valid reasoning" (5.267).

Guided by these methodological reflections, Peirce constructs his theory of mental activity. He starts with the externalist convictions that reasoning is an activity carried out in language, and that language is fundamentally a system of public signs. Moreover, since language use is not merely the vehicle of thought but the paradigmatic instance of it, it seems reasonable to view thinking as sign manipulation, thought as a sign process, and the structure of thought as the formal principles of this sign process. Accordingly, in explicating the structure of thought Peirce draws on both his general theory of signs and the principles of logic.[1]

Language and thought essentially involve representations of the world, one thing standing for another, one thing being a sign of another. Hence any understanding of the function of language and thought would seem to involve fundamentally a general account of representation, a general theory of signs. This is where Peirce starts. He defines a sign as "anything which determines something else (its *interpretant*) to refer to an object to which it itself refers (its *object*) in the same way, the interpretant becoming in turn a sign, and so on *ad infinitum*" (2.303). As it stands the definition is maximally general, ranging over both natural and conventional signs and including no essential reference even to minds.

In contrast to those who construe the signification relation as a simple dyadic relation between a sign and its object, Peirce sees it as irreducibly triadic. Something functions as a sign of a given object to something else: A is a sign of *B* to C. Although actual signification involves interpretation, it is important to note that in the triadic sign relation itself, the third relata, the C—that to which A is a sign of *B*—is not a mind but rather another sign which 'interprets' A as a sign of *B*. To make clear this distinction between an interpreter and an interpretant, we can attend to another of Peirce's definitions of "sign":

> A sign is anything which is related to a second thing, its object, in respect to a Quality, in such a way as to bring a thing, its interpretant, into relation to the same object, and that in such a way as to bring a fourth into relation to that object in the same form *ad infinitum*. (2.92)

Signs, then, are not special kinds of things but rather anything which functions in a certain special way, namely, to relate a third thing, itself a sign, to the object of which the original sign is a sign. Not only can anything function as a sign, but the very same sign vehicle can signify differently in different contexts. Signs are signs only in virtue of their playing a certain functional role in a process wherein signs interpret other signs ad infinitum. Logically speaking, there is no first sign nor is there any last sign; interpretation goes all the way down and all the way up.

The paradigm case of a sign process for our purposes is communicative language, for it is these sign processes which are instances of thought and are to be our models for understanding all mental activity. Signs on this level are to be understood functionally and holistically. Terms are specified in virtue of their roles in sentences, and sentences in virtue of their roles in argument. The concrete phenomenon we are dealing with is the *system* of language through which we interpret and refer to our world; this is analyzed as being made up of sentences which in turn are made up of terms. But from the fact that meaningful language can be analyzed thusly it does not follow that it is built up atomistically from independently meaningful units. In a functional analysis the whole is prior to the parts, the whole in this case being the argumentatively structural linguistic system "in which a term is a rudimentary proposition and a proposition is, in its turn, a rudimentary argumentation" (2.344).

The individual signs which analytically make up this system can be seen to be of different kinds depending on the level of analysis. In addition to the simple distinction between terms, sentences and arguments, Peirce's most important distinction is that between icons, indices, and symbols, a distinction based on the ways in which various signs are related to their objects. Icons are those signs which represent their objects in virtue of a qualitative resemblance to them (paintings, maps); indices represent their objects in virtue of some dyadic existential relation, such as causality or indication (weather vanes, directional arrows); symbols represent their objects only in terms of some general conventions or rules (nautical flags, Morse code). Since there is a dimension of conventionality in any genuine sign relation (rules for 'reading' maps or arrows), there are no pure icons or indices in a sign system. Nevertheless, there is an important and irreducible difference between that in virtue of which each of these three kinds of signs functions to represent its object.

This distinction is important for our purposes because Peirce thinks that all three kinds of signs are necessary in any language adequate for making meaningful assertions about reality:

> I have taken pains to make my distinction of icons, indices and tokens [replicas of symbols] clear, in order to enunciate this proposition: in a perfect system of logical notation signs of these several kinds must all be employed. Without tokens there would be no generality in the statements, for they are the only general signs; and generality is essential to reasoning.... But tokens alone

> do not state what is the subject of discourse; and this can, in fact, not be described in general terms; it can only be indicated. The actual world cannot be distinguished from the world of imagination by any description. Hence the need of pronouns and indices, and the more complicated the subject the greater the need of them.... With these two kinds of sign alone any proposition can be expressed; but it cannot be reasoned upon, for reasoning consists in the observation that where certain relations subsist certain others are found, and it accordingly requires the exhibition of the relations reasoned within an icon.... All deductive reasoning, even simple syllogism, involves an element of observation; namely, deduction consists in constructing an icon or diagram the relations of whose parts shall present a complete analogy with those of the parts of the object of reasoning, of experimenting upon this image within the imagination, and of observing the result so as to discover unnoticed and hidden relations among the parts. (3.363)

Representing the world, in a sense of "represent" essentially involving understanding, is a matter of subsuming its items under general categories. General terms are symbols that ascribe properties to things, which properties could be ascribed to other objects using the same sign. In short, they are predicates, and the generality involved in predication is clearly essential to representation and understanding—but it is not sufficient. There must, in addition, be items in the language that simply identify the objects of the various predications, and these items are the indices. Examples of indices are demonstratives, pronouns, proper names, and quantifiers. It is by means of these indexical terms that objects are identified or pinned down, to which we then ascribe properties. Thirdly, if this language is to be used for reasoning, symbols and indices are still not sufficient: "The arrangements of the words in the sentence, for instance, must serve as *Icons*, in order that the sentence may be understood" (4.544). The sentence's relation to other sentences which it interprets and still others which interpret it must be perspicuous if its role is to be discerned; and the kind of perspicuity involved is schematic or diagrammatic. This necessitates icons. All three kinds of signs, then, are necessary for any language rich enough to be a vehicle for, or an instance of, thought. As a summary statement for this series of points, we could say that any language rich enough to sustain reasoning about the world must have symbolic, indexical, and iconic dimensions.

These signs in a language are not representative atomistically but only as parts of a sign process wherein they interpret earlier signs and are in turn interpreted by later ones. Moreover, on the hypothesis Peirce is exploring, the sign process has a determinate structure, inasmuch as the ways in which signs determine other signs fall into one of three irreducible patterns, namely, those of deductive inference, inductive inference, or abductive inference.

Some sign processes have the following structure: the facts as represented in the earlier sign vehicles are such as to compel us to represent the facts in the later way. This is the deductive structure. Secondly, other sign processes are less 'deterministic': the facts as represented in the earlier sign vehicles are not such as to compel the later representations but 'invite' them. The earlier representations may depict the objects as having properties they clearly would have if they were structured as represented in the later sign vehicle. This is the abductive structure. Or, the facts as represented in the earlier sign vehicles again do not compel the later ones, but the later ones are just extrapolations from the earlier ones. This is the inductive structure (2.96).

It is Peirce's contention that all sign processes wherein signs are interpreted by subsequent signs of the same objects fall into one of these three patterns, namely, one of the three forms of inference. The earlier sign vehicles can be seen as premises leading to the later conclusions either determinately, by suggestion, or by extension. He specifies this even further in his claim that all sign processes exhibit one of the forms of 'valid inference,' and explains the obvious cases of invalid reasoning as instances of false premises, confusion of rules, or just cases of extremely impoverished abductions or inductions (5.282).

The sign process so conceived is a process without logical end and without logical beginning. Inasmuch as every sign, to be a sign, must potentially give rise to an interpretant which is itself a sign, obviously there can be no such thing as a last sign. The sign process can be and often is interrupted, but in such instances it is not completed, logically speaking. We can, however, conceive of its final completion in terms of the limit of the continuous series of which it is a fragment. The endlessness in the other direction is more elusive. Although there undoubtedly was a time before which there were no signs in this rich sense, it does not follow from this that there was a single first sign unpreceded by any other sign. It is Peirce's view that every sign has as its immediate object another sign which it interprets (and through which it refers to its external dynamical object), such that there is a potential infinity not only of subsequent signs but of precedent signs. What makes this picture entertainable is his conception of the beginning of signification not as a discrete moment but as a process which can only be represented as a dense series: "Cognitions ... thus reach us by this infinite series of inductions and hypotheses (which though infinite *a parte ante logice*, is yet one continuous process not without a beginning *in time*" (5.311).

Peirce's views about these general features of signs are instantiated in his view of language. He thinks of language not primarily as a formal system but as a practice, i.e., a dialogue or conversation. Although this dialogue may be interrupted, it is logically speaking an endless conversation given the finitude of any perspective and the inexhaustibility of our concrete world. It is also the case that language doesn't have an absolutedly discrete beginning. Although undoubtedly there was a time before which linguistic practice didn't obtain, the practice didn't begin with a discrete 'first word'. Words are words only in virtue of their roles in a general system of linguistic communication; so to have the use of a word is to have access to many words and to the rules that govern their use. This doesn't entail that linguistic practice can't begin but only that it can't begin discretely or atomistically, but only as a systemic process. It is a form of behavior that itself grows and develops and into which we are gradually introduced.

It is this view of language and sign behavior which is extended analogically in Peirce's account of thought as 'inner speech'. Thought is construed as mental words, and thinking in terms of mental language; and this mental language is to be understood in terms of categories appropriated from our account of public language. All thought is in signs because most fundamentally "every thought is a sign" (5.253), and this sign system, inner language, which is thought, is not to be understood atomistically and statically but as a holistic and dynamic communicative system.

What this means for Peirce is that all thought is conversational or dialogic in form: "thinking always proceeds in the form of a dialogue—a dialogue between different phases of the ego" (4.6). Even our private thoughts have this communicative structure:

> A person is not absolutely an individual. His thoughts are what he is "saying to himself," that is, is saying to that other self that is coming into life in the flow of time. When one reasons it is that critical self that one is trying to persuade.... (5.421)

Peirce goes on to make the strong claim that it is not merely a contingent fact but rather a conceptual truth that thought has this dialogic structure: "It is not merely a fact of human Psychology, but a necessity of Logic, that every logical evolution of thought should be dialogic" (4.551).

When he speaks of thought as inner language, the notion of "language" here has to be construed broadly. Although to think in words is to think in particular words, the thought is not identical with any of the particulars: "One selfsame thought may be carried upon the vehicle of English, German,

Greek or Gaelic; in diagrams, or in equations, or in graphs; ... yet that the thought should have *some* possible expression for some possible interpreter is the very essence of its being" (4.6). It is important to note that language so construed contains much more than words:

> The majority of men commune with themselves in words. The physicist, however, thinks of experimenting, of doing something and awaiting the result. The artist, again, thinks about pictures and visual images, and largely in pictured bits; while the musician thinks about, and in, tones. Finally, the mathematician clothes his thought in mental diagrams which exhibit regularities and analogies of abstract forms.... (*CN* 3.258–59).

Just as our ordinary language narrowly construed is in practice enriched by the inclusion of other sign systems, so too with the language of thought. That it includes diagrams, formulae, and graphs is not surprising for someone whose primary instance of knowledge is science, but its capacities are as broad as the fields of human accomplishment.

Since this account of thought is achieved by the analogical extension of his account of language and signs, we would expect some of the other features of the model to carry over. His fallibilism and his limit conception of truth are quite naturally suggested. Our thoughts always need to be interpreted by and supplemented by subsequent thoughts, and this process of understanding is never complete. We can, however, conceive of its completion, conceive of arriving at the truth, in terms of the limit of the series of which our present thoughts are fragments. The de facto endless conversation can be conceived of as having a natural culmination.

To this point I have been focusing on the kind of mental activity (acts of understanding) that would seem to be the most easily assimilable to Peirce's model. Maybe one can give a plausible account of *thinking* on this model, but mental activity includes much more than thinking. What would it even mean to reduce sensation, emotion, attention, and "all mental action to the formula of valid reasoning" (5.267)? The structure of the thought process might best be understood by an analogical extension inward of the logical structure of language, but what would it mean to extend this account to the structure of the rest of our mental life?

True to his methodological convictions, Peirce feels that the presumption should be in favor of such an extension: "Every phenomenon of our mental life is more or less like cognition; every emotion, every burst of passion, every exercise of will, is like cognition" (1.376). This being the case, the structure of cognition, i.e., the inferential structure, should be presumed to be embodied in, and thus be explanatory of, these other instances of our mental life.

The first point that he wants to make is that our knowledge of our sensations, emotions, and intentions is not via some internal scanning of mental states but rather a mode of 'returning reason' providing commentary on some features of public objects. Our knowledge of our sensation of red, for example, would be a matter of recognizing that we are undergoing that sensation which we normally undergo in the presence of red objects: "That knowledge would in fact be an inference from redness as a predicate of something external" (5.245). That same analysis is extended to our knowledge of our emotions and other mental states. To become aware of myself as angry is to recognize that I am undergoing that kind of experience which I normally have in the presence of objects I revile. Peirce maintains that reflection will show that an individual's anger "consists in his saying to himself 'this is abominable, etc.', and that it is a mark of returning reason to say, 'I am angry'" (5.247). The same general strategy is pursued to account for our access to our sense of beauty, moral sensibilities, and intentions. In this way Peirce extends his model of our access to the mental to performances other than thought. However, whatever be the case with regard to our *knowledge* of these mental states, it remains to be seen whether the states themselves can be conceived of on the model of drawing inferences. How can this model explain the structure of sensations, emotions, and other mental states, even granted that it functions as a plausible account of our knowledge of such states?

Peirce starts with sensation and assimilates the structure of sensation itself to the model of abductive inference. A sensation arises as a psychological state which unconsciously reduces a manifold of impressions to an experiential unit. For example, "the sensation of a particular kind of sound arises in consequence of impressions upon various nerves of the ear being combined in a particular way and following one another with a certain rapidity"; "the sensation of color depends upon impressions upon the eye following one another in a regular manner and with a certain rapidity"; and "the sensation of beauty arises upon a manifold of other impressions" (5.291). Peirce's suggestion is that we think of the sensation itself as a unifying predicate which enables us to synthesize discrete information into a simple characterization, just as by abduction we unify discrete data under a general theory: "A sensation is a simple predicate taken in place of a complex predicate; in other words, it fulfills the function of an hypothesis" (5.291). Obviously this account involves a commitment to unconscious information processing and unconscious inference, but Peirce already acknowledged as much in his critique of Descartes.

Emotions are construed as exhibiting the same structure. Certain emotions arise in us when our attention is drawn to complex and hard-to-understand circumstances. Anxiety on the one hand, and wonder on the other, arise in us when in the face of incomprehensible data we are either uncertain about our fate or curious about their ultimate intelligibility. In each case the emotion is really "a simple predicate substituted by an operation of the mind for a highly complicated predicate" (5.292). What distinguishes sensations and emotions from thoughts is not the structure of the mental process but the relative prominence of the material quality (the felt immediacy) of the mental action.

Attention as a mental action seems to be quite different. Out of the normal flow of experience we do attend to some phenomena more than others, and, as a result, these attended to phenomena produce a greater effect on memory and subsequent thought. Attention is aroused when the same phenomenon presents itself repeatedly on many occasions: "We see that *A* has a certain character, that *B* has the same, C has the same; and this excites our attention, so that we say '*these* have this character'; thus attention is an act of induction" (5.296). Basically a habit is formed, and the structure of the habit formation is the structure of induction—in this case the simplest kind of enumerative induction. Having sketched what an account of these three different kinds of mental action would look like, Peirce concludes: "We have thus seen that every sort of modification of consciousness—attention, sensation and understanding—is an inference" (5.298).

This is a general sketch of Peirce's 'unified theory' of mental activity. On this account the items are understood primarily in functional terms, and the explanation of the mental is quite independent of specific instantiations. Hence, even if substantially correct, as an account of *our* mental life it is essentially incomplete.

b. The Qualitative Dimension of Mental Activity

There is more to our mental life than its structure; there is the qualitative dimension that is characterized in terms of feeling, subjectivity, and consciousness. Our mental life has matter as well as form. Two preliminary points need to be made about this. First, Peirce's denial of introspection was not a denial of an inner life of feeling or subjectivity but only the denial of a certain direct manner of access to it. Secondly, what might with some liberty be called his functionalist account of mind cannot be construed as a rejection of this dimension of subjectivity or feeling (which he categorized as 'consciousness'), because he clearly distinguished mental phenomena from conscious phenomena: "Mind is not consciousness nor proportionate in any way to consciousness" (7.365). There is both non-mental consciousness and unconscious mind. Mental phenomena are complex functional states, whereas consciousness is a simple thing, "nothing but feeling" (7.365).[2]

The general picture he seems to have in mind is a version of a dual aspect theory, where feeling or consciousness is the internal dimension of certain behavioral interactions with our environment, many of which interactions are mentally structured:

> What is meant by consciousness is really in itself nothing but feeling. Gay and Hartly were quite right about that; and though there may be and probably is, something of the general nature of feeling almost everywhere, yet feeling in any ascertainable degree is a mere property of protoplasm, perhaps only of nerve matter. Now it so happens that biological organisms and especially a nervous system are favorably conditioned for exhibiting the phenomena of mind also; and therefore it is not surprising that mind and feeling should be confounded. But I do not believe that psychology can be set to rights until the importance of Hartmann's argument is acknowledged, and it is seen that feeling is nothing but the inward aspect of things, while mind on the contrary is essentially an external phenomenon. (7.364)

Peirce here tips his hat to one of his more speculative cosmological views, namely, "that there probably is something of the general nature of feeling almost everywhere," but claims that, in the sense in which we are concerned with it, consciousness is a property of the nervous systems of biological organisms: "We know that the protoplasmic content of every nerve-cell has its active and passive conditions, and argument is unnecessary to show that feeling, or immediate consciousness, arises in an active state of nerve-cells; ... though we cannot say that every nerve-cell in its active condition has feeling (which we cannot deny, however) there is scarce room to doubt that the activity of nerve-cells is the main physiological requisite for consciousness" (1.386).

In a passage that clearly underscores his strong claim that a nervous system is a necessary condition for a developed consciousness, Peirce draws a straightforward but somewhat surprising consequence:

> Since God, in His essential character of Ens Necessarium, is a disembodied spirit, and since there is a strong reason to hold that what we call consciousness is either merely the general sensation of the brain or some part of it, or at all events some visceral or bodily sensation, God probably has no consciousness. Most of us are in the habit of thinking that consciousness and psychic life are the same thing and otherwise to overrate the functions of consciousness. (6.489)

God is thought of as intelligent but not as conscious, whereas some lower animals are thought of as having consciousness but lacking intelligence. Hence, "mind" and "consciousness" pick out quite different features of objects. In the familiar cases, however, since a certain level of organic complexity is involved in both mental behavior and consciousness, one is mistakenly identified with the other. But in those cases where they go together (and there are cases where they do not) consciousness is related to the mental as the inner aspect is to the outer.

Peirce's most direct (but hardly straightforward!) definition of consciousness contains another very important feature of his view, i.e., that his view is not at all epiphenomenalist:

> Consciousness may be defined as the congeries of non-relative predicates, varying greatly in quality and in intensity, which are symptomatic of the interaction of the outer world—the world of those causes that are exceedingly compulsive upon the modes of consciousness, with general disturbance sometimes amounting to shock, and are acted upon only slightly, and only by a special kind of effort, muscular effort—and of the inner world, apparently derived from the outer, and amenable to direct effort of various kinds with feeble reactions; the interaction of these two worlds chiefly consisting of a direct action of the outer world in the inner and an indirect

> action of the inner world upon the outer through the operation of habits. If this be a correct account of consciousness, i.e., of the congeries of feelings, it seems to me that it exercises a real function in self-control, since without it, or at least without that of which it is symptomatic, the resolves and exercises of the inner world could not affect the real determinations and habits of the outer world. (5.493)

Consciousness exercises a real function in self-control, but indirectly "through the operation of habits." What Peirce has in mind here is less than clear, and the only help he gives us is a sustained metaphor. He compares consciousness to a bottomless lake in which ideas are suspended at different depths. There is a gravitational force operating such that the deeper the ideas, the more effort in the form of energy of attention is required to bring them to the surface. Moreover, these ideas are attracted to each other via associational habits of continuity and resemblance. By a certain effort of attention we can bring certain ideas (and those associated with them) close to the surface, where they can then function in guiding our interaction with the world (7.553–55). Obviously much more would have to be said before the adequacy of such an account could even be assessed.[3]

Whatever the details of this positive account, the fact that it is intended as non-epiphenomenalist is quite clear. Peirce saw epiphenomenalism as a by-product of the attempt to offer a deterministic account of all events in terms of mechanical laws, and he even acknowledged that this attempt was "good scientific procedure ... which had to be tried and persevered in until it was thoroughly exploded" (5.64). But mechanism in his mind was now seen to be exploded, and one of the main objections to it was that it reduced consciousness to a "mere illusory aspect of a material system" (6.61). It was part of his argument for the statistical interpretation of scientific laws that such would allow conceptual space for consciousness to play a real causal role in the self-control of developed organisms.

This insistence on the qualitative dimension of mental activity reemerges in Peirce's answer to a very unusual question prompted by his structural account of mind. If mental events are basically signs and man is a series of such signs (in a sense to be explained shortly), what exactly distinguishes man from a word or a series of words? Part of the answer is in terms of consciousness:

> It may be said that man is conscious, while a word is not. But consciousness is a very vague term. It may mean that emotion which accompanies the reflection that we have animal life. This is a consciousness which is dimmed when animal life is at its ebb in old age or sleep, but which is not dimmed when spiritual life is at its ebb; which is more lively the better *animal* a man is, but which is not so the better *man* he is. We do not attribute this sensation to words, because we have reason to believe that it is dependent on the possession of an animal body. (5.313)

That there should even be a comparison between the nature of a man and a word is as interesting as the difference adduced. The comparison will be discussed in greater detail in the next section, but it is introduced here as an instance of his inner aspect view.

Consciousness as Peirce defines it is a matter of simple feeling, but this is not at all true for self-consciousness. Self-consciousness on his account is a much more complicated matter. For him, full self-consciousness is a cognitive state and as such is inferential in nature: "One is immediately conscious of his feelings, no doubt; but not that they are feelings of an *ego*; the *self* is only inferred" (5.462). From his earliest papers he had argued that our awareness of ourselves as such was a cognitive achievement accomplished over time via inferences from the rudimentary experiences of ignorance and error. Self-consciousness in this rich sense is neither primitive nor simple but is "the result of inference" (5.237).

This full self-consciousness has its phenomenological origin in a rudimentary polarity (secondness) involved in some experiences:

> But the consciousness of compulsion in sensation as well as the consciousness of willing necessarily involves self-consciousness and also the consciousness of some exterior force. The self and the not-self are separated in this sort of consciousness. The sense of reaction or struggle between self and another is what this consciousness consists in. (7.543)

It is this primitive feeling of 'otherness' or 'over-againstness' that gives rise to the chain of inferences that ultimately results in the cognitive state that is knowledge of the self.[4]

Such is Peirce's account of the two aspects of mental activity: a formal or functional account of mental states together with an insistence on the distinct phenomenon of consciousness understood as feeling or subjectivity. The account to this point, however, has been somewhat closely moored to the specific phenomena to be explained, namely, the elements of our mental life. But Peirce, as a product of the late nineteenth century, has a more robustly speculative side, and it is to these broadly metaphysical ruminations that I will now turn.

2. The Metaphysical Status of Mind and Reality

Man's identity consists neither in his consciousness nor in his will (which he shares with animals) but in the level of his mental capacity, a level characterized by his deliberate use of signs:

> Now the intelligent control of thinking takes place by thinking about thought. All thinking is by signs; and the brutes use signs. But they perhaps rarely think of them as signs. To do so is manifestly a second step in the use of language. Brutes use language, and seem to exercise some

> little control over it. But they certainly do not carry this control to anything like the same grade that we do. They do not criticize their thought logically. (5.534)

We are conscious of our use of signs, and we reflect on this use and thereby are able to exercise a level of control over our thoughts that informs our behavior with intelligence and purpose. From Peirce's perspective it is even misleading to say that what is distinctive about us is this controlled use of signs, because this suggests an 'us' which is distinct from the sign process. But given his view that it follows from his critique of incognizables that "the phenomenal manifestation of the substance is the substance" (5.313), Peirce concludes that "the sign which the man uses *is* the man himself, for ... the fact that every thought is a sign, taken in conjunction with the fact that life is a train of thought, proves that man is a sign" (5.314). The proximate conclusion, i.e., that man's life is a sign, gives way to the more sweeping claim that man himself is a sign, on the assumption that "the manifestation of the substance is the substance."

In what might be thought of as an intellectualized version of the 'stream-of-consciousness' view of man, then, Peirce construes man in his distinctiveness as quite literally a dynamic system of signs or, by way of abbreviation, as a complex sign. That Peirce takes this characterization of man quite literally is manifest in his serious consideration of this question: what exactly is the difference between that complex sign which is man and other more familiar complex signs such as books, sentences, or words? Given that his first response to the question—"What is man?"—is that man is a symbol, he goes on to acknowledge that "to find a more specific answer we should compare man with some other symbol" (7.583). A man and a word are structurally of the same kind; in what does their difference consist?

The first point that Peirce makes is that the most obvious differences between a man and a word are not functions of their status as symbols but of their different embodiments.

They are both formal realities, but "the body of man is a wonderful mechanism, that of the word nothing but a line of chalk" (7.584). Many of the more obvious differences are grounded in this difference of embodiment. The most basic difference is that man is conscious and the word is not; but as we have already seen, Peirce traces this basic difference to "the possession of an animal body" (7.585). We should not expect all mental realities—even God—to be characterized by consciousness. Similarly with the obvious difference that man has perceptions, i.e., he is affected by objects, in ways in which words do not. This too depends on "having an animal organism" (7.587). Again, man has a moral nature whereas words clearly do not. But since morality has to do with action and dispositions to act, this difference is also tied to the fact that words are "physiologically incapacitated to act [so] we should not consider this as a separate point of distinction" (7.588). On the basis of the these reflections, Peirce remarks that "you see that remote and dissimilar as the word and the man appear, it is exceedingly difficult to state any essential difference between them except a physiological one" (7.588). These obvious differences don't seem to bear on the essence of man, which is formal rather than material.

Peirce makes an interesting observation with regard to the claim that man procreates whereas words do not. If this is intended not simply as another point about physiology but about mentality, Peirce denies it:

> Has the word any such relation as that of father and son? If I write "Let *kax* denote a gas furnace," this sentence is a symbol which is creating another within itself. Here we have a certain analogy with paternity; just as much and no more as when an author speaks of his writings as his offspring—an expression which should not be regarded as metaphorical but merely as general. (7.590)

Peirce distinguishes here between metaphor and analogy, and insists that his comparison of a man and a word is the latter rather than the former. Moreover, he sees analogy, understood as a "broad comparison on the ground of characters of a formal and highly abstract kind" (7.590), as the medium of sound metaphysical speculation.

Man is not one among the animals; he has an essence that transcends his embodiment:

> There is a miserable material and barbarian notion according to which a man cannot be in two places at once; as though he were a *thing*! A word may be in several places at once, Six Six, because its essence is spiritual; and I believe that man is no whit inferior to the word in this respect. Each man has an identity which transcends the mere animal;—an essence, a *meaning* subtile as it may be. (7.591)

A man is not a mere thing. He is no more identical with his animal body than a word is identical with a given inscription. Inasmuch as my thoughts and sentiments—which *are* me—are not confined to a unique time and place, I am a transcendent reality.

Peirce characterizes this essence of man in terms of information:

> This essence of which I speak is not the whole soul of man; it is only his core which caries with it all the information which constitutes the development of man, his total feelings, intentions, thoughts. (7.592)

The essence of man, then, is purely formal. To the degree that a person's thoughts and sentiments are true and authentic, he can achieve a kind of immortality. Peirce acknowledges that "this is an immortality very different from what most people hope for, although it does not conflict with the latter" (7.593). It is an immortality "which depends on man's being a true symbol" (7.594). The sum total of information is the essence of man, and his immortality is a function of the status of this information: "If the idea is true, he lives forever; if false, his individual soul has but a contingent existence" (7.595).

If we add to this 'semiotic theory of man' Peirce's claim that "signs are the only things with which a human being can, without derogation, consent to have any transaction" (6.344), it would seem that Peirce is faced with a quandary. If the physical universe is radically distinct from the semiotic order, he would seem to have an interaction problem not unlike that of Descartes; if this is to be avoided, he would seem obliged to pursue the initially unpromising path of construing the entire physical world along semiotic lines.

Peirce avoids the Cartesian problems of interaction by opting for a general metaphysical monism:

> It would be a mistake to conceive of the psychical and the physical aspects of matter as two aspects absolutely distinct. Viewing a thing from the outside, considering its relations of action and reaction with other things, it appears as matter. Viewing it from the inside, looking at its immediate character as feeling, it appears as consciousness. These two views are combined when we remember that mechanical laws are nothing but acquired habits, like all the regularities of mind, including the tendency to take habits itself; and that this action of habit is nothing but generalization, and generalization is nothing but the spreading of feeling. (6.268)

There are two separate points interwoven in this passage, but the overall intent is clear. First the mental world and the physical world are continuous rather than radically distinct, so the stark problem of interaction doesn't exist. The physical world is structurally of a piece with the mental, since mechanical laws are limit cases of mental laws. Secondly, consciousness is not tied to a radically different kind of entity from the physical but really is just the inner aspect of some complex physical systems. So neither mentality nor consciousness requires an entity different in kind from the physical order. Monism is the preferred option. The 'reductionism' involved in Peirce's monism, however, is in the opposite direction from that most frequently encountered.

His general metaphysical account is fashioned on the presumption of continuity plus an argued preference for idealism over materialism. Peirce is quite explicit about the motivation behind his metaphysical options. Given his rejection of dualism in favor of some form of metaphysical monism, he formulates the available options in terms of the reducibility or irreducibility of laws. Materialism is the view that physical laws are basic and psychical laws are derived; idealism is the view that psychical laws are basic and physical laws derived; whereas what he calls neutralism is the view that physical and psychical laws are irreducible. He invokes Ockham's razor against the third alternative and then makes the case for idealism over materialism.

He formulates materialism in such a way as to include mechanism or determinism, and then argues against it on several fronts. First, mechanism has shown itself to be inadequate as a fundamental account of scientific laws. Given an element of pure chance in the universe, a statistical interpretation of scientific laws seems to be mandated. Secondly, it seems impossible to explain the phenomena of growth and complexification on the assumption of mechanism. And, finally, mechanism seems completely unable to account for the emergence of consciousness and self-control in the universe. Obviously, this line of argument is only as good as the identification of materialism and mechanism.[5]

The remaining alternative is idealism, and this is the option that Peirce defends:

> The one intelligible theory of the universe is that of objective idealism, that matter is effete mind, inveterate habits becoming physical laws. (6.25)

On this general metaphysical picture, the law-like character of events is understood on the model of psychological laws, where 'habit formation' rather than mechanical determinism is the guiding intuition. Material interactions can be understood as limit cases of mental interactions where spontaneity and creativity are at their lowest ebb: "Physical events are but degraded or undeveloped forms of psychical events" (6.264).

Moreover, there is a dynamic dimension to this metaphysical account, whereby from an original chaos of chance our structured universe emerges by an evolutionary process of habit-taking: "The hypothesis suggested by the present writer is that all laws are the result of evolution; that underlying all other laws is the only tendency which can grow by its own virtue; the tendency of all things to take habits" (6.101). The fundamental principle of structure in the universe, i.e., the tendency to habit-taking, has the form of a psychological law and it is through the operation of this tendency that other structures develop.

In what Peirce terms his "guess at the riddle of the sphinx" (1.410), he sketches a cosmological story:

> Pairs of states will also begin to take habits, and thus each state having different habits with reference to different other states will give rise to bundles of habits, which will be substances. Some of these states will chance to take habits of persistency, and will get to be less and less liable to disappear; while those that fail to take such habits will fall out of existence. Thus substances will get to be permanent.
>
> In fact, habits, from the mode of their formation, necessarily consist in the permanence of some relation, and therefore, on this theory, each law of nature would consist in some permanence, such as the permanence of mass, momentum and energy. In this respect, the theory suits the facts admirably. (1.414–15)

Peirce, of course, recognizes the tenuousness of such cosmological speculation, and indicates what would be necessary to bring it in line with the canons of scientific reasoning: "We must show that there is some method of deducing the characters of the laws which could result in this way by the action of habit-taking on purely fortuitous occurrences, and a method of ascertaining whether such characters belong to the actual laws of nature" (1.410).

Whatever its warrant, this kind of objective idealism allows Peirce to generalize his theory of signs into a semiotic view of reality. Not only are signs constitutive of mental behavior but, given his mentalistic characterization of laws, of all interactions whatsoever: "All this universe is perfused with signs, if not composed exclusively of signs" (5.448 n.1 referring to 4.539). As the final chapter to this story, Peirce adds the theological speculation that "the universe is a vast representamen, a great symbol of God's purpose, working out its conclusions in living realities" (5.119).

V. *Conclusion*

Any concrete historical life is much more than any representation. This truism takes on heightened significance if the life and vision in question are those of a multi-faceted creative genius whose intellectual ruminations transcended traditional categories and established conventions. My Peirce of necessity represents a selection, and there are many facets of his thought that I have totally ignored and more still that I have underemphasized. However, selection is not falsification, and it is my contention that the facets upon which I have focused—his views on science, knowledge, and mind—are at the absolute center of Peirce's philosophical vision. But I'm sure all interpreters feel this way about their interpretations. If readers feel that in virtue of my selection and emphasis some important misrepresentation has been perpetrated, I can only hope that my effort moves them to return to that remarkably fecund set of texts for the articulation of their Peirce in order to redress the balance.

Furthermore, philosophical interpretations are always from the perspective of the interpreter and are in part motivated by present philosophical concerns. It is his potential participation in our philosophical conversations, either by way of criticism or supplementation, that makes the historical figure philosophically interesting. This being the case, in this conclusion I would like to suggest several points at which Peirce's vision can be profitably introduced into our present philosophical reflections. His philosophical orientation was defined by concerns that still motivate ours, and on several issues his perspective really is a suppressed alternative in present conversations. Once introduced, his perspective could lead our discussions in new directions. I will start at the general level with his implicit conception of the interrelation of the issues of science, knowledge, and mind and then move on to more specific points under each category.

In the twentieth century philosophy of science, epistemology, and the philosophy of mind have been for the most part pursued independently. Each has had its own practitioners and its own literature, and even when addressing the same general issues the literatures rarely overlap. The theory of knowledge has been pursued as if science isn't knowledge, and philosophy of mind developed as if scientific investigation were not a paradigmatic mental activity. If one adds to this level of fragmentation the general development of philosophy quite independent of science, the result is a multi-dimensional compartmentalization of philosophical investigation, which is not only artificial but militates against that kind of cross-fertilization of thought that in the past has been the engine of creative advance.

Philosophical reflection cannot be rigorously compartmentalized. The questions quite naturally overflow the boundaries so it seems only appropriate that the attempts to answer them reflect a similar openness to different levels of information. The parts of philosophy are like parts of an organic system. Specialization is not inappropriate, but if one is really going to understand a given part one can't lose sight of its relations to the other parts and its overall function in the whole. Moreover, it would seem only natural that philosophical reflection be carried on in relation to other reasoned attempts to understand the very same world. Natural science, social science, history, and literature reveal facets of our world that philosophical reflection can ignore only at its own peril. One need not be a 'grand synthesizer' to recognize the conventionality of boundaries and the appropriateness of being quite generally informed about the domain under investigation.

Recently, this artificial compartmentalization has begun to break down under different pressures from two very different quarters. On the one hand, there is emerging what might be called a 'new positivism', which locates the answers to philosophical questions straightforwardly in the specialized sciences, e.g., neurophysiology, evolutionary biology, or cognitive science. On this construal there is no level of reflection left over after the special sciences complete their tasks. On the other hand, there is a kind of 'new criticism' which views the traditional problems of both philosophy and science as mere historical constructs that have to be overcome: our relation to the world is 'narrative' all the way down with no particular kind of narrative having privileged epistemic status. This second tendency is both anti-scientific and anti-rational in its orientation. In my judgment both of these reactions to traditional philosophizing—both of which not infrequently go under the label "pragmatism"—are fundamentally misguided and not at all in the spirit of classical pragmatism.

Historical pragmatism, particularly in the person of Charles S. Peirce, did transcend the compartmentalization and abstraction that ultimately proved so stultifying to philosophy, but in a way that was neither reductionistic nor anti-rational. Science was regarded as the paradigmatic instance of knowing, but in such a way that there was not only room for, but also a real need of, genuine philosophical reflection. Secondly, Peirce's critique of traditional construals of both philosophy and science was not at all anti-rational but rather a stage in the attempt to grasp concretely the true natures of philosophy, science, and rationality. Peirce saw philosophy of science, epistemology, and philosophy of mind not only as intimately related but as inextricably tied to concrete developments in science. His response to the dead ends in traditional philosophy was a genuine reformation rather than an outright rejection, and it is this kind of 'pragmatism' that seems to me to be the appropriate direction for future philosophizing. Hence, this overview of the thought of Charles S. Peirce has been intended not merely as an attempt to understand America's most enigmatic and creative philosophical genius, but also to serve as a suggestive model for ways in which our current philosophical hurdles might be overcome. In the remainder of this conclusion I intend to focus on several specific points in Peirce's philosophy of science, epistemology, and philosophy of mind that directly relate in suggestive and constructive ways to central discussions in contemporary philosophy.

Many of the issues on Peirce's agenda in the philosophy of science are still on center stage today. I will focus on three: first, the social and historical nature of science; secondly, the identity and preeminence of scientific method; and thirdly, the issue of scientific realism. It is my contention that Peirce's views are not only relevant to these contemporary discussions but would in several cases be a corrective force to present tendencies.

The sociology of science has had a two-stage development in twentieth century thought. It began with the simple recognition of the fact that since science viewed concretely is a kind of cognitive activity both social and historical in nature, like all other social phenomena it invites sociological analysis. As a cultural institution it has a history, an internal social structure, and external relationships to other social phenomena. The sociology of science emerged as the enterprise attending to these features of science. In its first stage, however, it left untouched the conception of science as a privileged kind of cognitive activity, not only thoroughly rational but the very paradigm of rationality. There were obviously social and historical dimensions to the behavior of scientists and to the institutionalization of science, but these weren't seen as determining its content or compromising its rationality. This early Mertonian tradition focused on the ethos of the scientific community analyzed as a system of controlled functional interactions, governed by norms and counter-norms which guided its rational progressive development. Studies of specialization, communication, stratification, and reward systems emerged, but a direct sociological account of scientific knowledge was avoided because the belief persisted that to the extent that knowledge is genuinely scientific, it is determined by the physical world and not the social world. The sociology of science was exempted from the sociology of knowledge.

The second stage in the sociology of science, epitomized in the Edinburgh School, was more thoroughgoing in its contextualization of science. Influenced by post-positivistic developments in the history and the philosophy of science, the epistemological barriers to a sociological analysis of scientific knowledge were lifted. The sociology of science took its place within the sociology of knowledge, and scientific knowledge was exhibited as influenced by external causal determinants much like other systems of belief and practice. Seen as an historically contingent social enterprise driven by various human interests, it might well have no distinctive claim to rationality or objectivity. The conclusions of science are construed as offering a conception of the physical world mediated by available cultural resources such that at the highest level of generality science should be viewed as an interpretive enterprise in which our picture of the world is socially constructed.

Peirce's contribution to this discussion can be seen to embody features of both the earlier and later sociological traditions; features which from our vantage point are supposed discordant seem to fit together in something like a coherent whole. On the one hand, he maintains that the aim of science is objective knowledge and truth, that there is a specifically characterizable scientific method, and that this method defines the paradigmatic case of rational cognitive behavior. On the other hand, he also maintains that scientific inquiry is informed by interests, structured by norms, and driven by certain ineliminable moral factors and social ideals. Added to these features are a causal-evolutionary account of scientific insight and an economic characterization of theory acceptance.

The point worth reflecting on is that on Peirce's account all these factors are 'intrinsic' to science and all are involved in an 'internal' account of scientific development. The social, historical, and moral factors are not seen to compromise or even qualify the objectivity of science, but to be partly constitutive of scientific rationality, scientific progress, and the realistic reach of science. This should call attention to the fact that one's stance on the issue of the rationality of scientific explanation and scientific change is as much a function of the scope of one's conception of rationality as it is of any particular characterization of scientific practice, and that one's stance on the issue of the realistic reach of science is as much a function of one's conception of reality as of any historical story of what scientists do. There is a symbiotic relationship between the articulation of defensible abstract notions of 'rationality' and 'reality' and a familiarity with the concrete development of human cognitive accomplishments.

Part of the explanation of Peirce's ability to satisfy himself that all of the aforementioned factors can be woven together into an internally coherent account of science would be in terms of his richer (or at least broader) conceptions of rationality and reasoning, conceptions which he gets in part from his reflections on the history of thought. Speaking approvingly of Lavoisier, Peirce identifies as one of his major accomplishments the articulation of "a new conception of reasoning as something which is to be done with one's eyes open, in manipulating real things instead of words and fancies" (5.363). Peirce himself continued this line of development and expanded the concepts of rationality and reasoning in the direction of further concreteness by including historical, social, institutional, and valuational factors as well. As the concept of rationality expands, the kinds of factors it is rational to take into account in accepting or deciding between scientific theories correspondingly expands; conversely, the kinds of 'extra-rational' factors contract. Moreover, this matter of definition of the key notions is not ad hoc or purely conventional. Illuminating concepts do not descend from a Platonic heaven but are fashioned through their own historical dialectic. This is particularly true of fundamental normative notions such as 'rational' and 'real', and Peirce would defend *his* formulation of such notions in terms of their ability to shed light on the history of and practices embodied in our cognitive enterprises. They emerge from the process they are endeavoring to illuminate.

The second set of issues in this area to which I want to call attention has to do specifically with scientific method. Peirce argues that there is such a structure as *the* scientific method and that it is cognitively privileged. Both are contested issues in our post-positivistic age. First things first: Hilary Putnam is among those who have argued that the appeal to 'the scientific method' is empty:

> My own view to be frank, is that there is no such thing as *the* scientific method. Case studies of particular theories in physics, biology, etc., have convinced me that no one paradigm can fit all the various inquiries that go under the name of science.[1]

Putnam's point is that while there are responsible and irresponsible ways of conducting inquiry, when one canvasses the variety of investigations that invoke the name of science, one is hard pressed to come up with some common characterization of method that is non-vacuous.

One need only call to mind extremes such as 'creation science' to realize that Peirce's point is not intended to encompass everything that goes under the name of science but only those inquiries that have a 'legitimate' claim to this designation. The question is, is there any way to parse this notion of legitimacy without begging the question? The seeds of a detailed "yes" answer can be found in Peirce's strategy of invoking historical, social, and pragmatic criteria to delimit the range of candidates, and then to attempt to draw from this select list meaningful generalizations about method. The point is that many kinds of cognitive enterprises have been initiated under the name of science but only some have proved to be socially successful. Only some have survived the test of time, in the sense that they have convinced a diverse and ever-widening expert community that certain issues have been settled, certain problematic situations resolved, and certain strategies effective. Physics, chemistry, biology, and some parts of psychology have shown themselves to be among the survivors; with regard to others the jury is still out; and some inquiries begun with enthusiasm have clearly fallen by the wayside.

Can meaningful generalizations be drawn from these paradigmatically successful sciences that would support a non-vacuous characterization of *the* scientific method? Peirce's suggestion—inference to the best explanation coupled with continual efforts at intersubjective empirical confirmation or falsification—while quite general, seems robust enough to demarcate genuine science from other kinds of inquiry. The fact that this characterization of science in terms of method may not be defensible against general skeptical attacks should not count against a view premised on the conviction that the Cartesian strategy was a fundamental mistake.

Given that this characterization of *the* scientific method is non-vacuous, the same historical, social, and pragmatic criteria can be invoked to make the case for its preeminence. The aim of cognition is to find out information about our world so that we may interact with it effectively. The general strategy that Peirce has characterized as scientific method, in contrast to other ways of proceeding, has established its credentials. The case can be made that it is only when our general puzzlements can be so stated that they can be dealt with in this way that we come up with answers that are in fact intersubjectively satisfying. This method, in contrast to the others, passes the relevant social test.

Finally, given that it does make sense to talk about *the* scientific method and that some case can be made for *its* preeminence, questions can be raised about its relation to history. Since Peirce does not make strong claims about present scientific theories but only about the culminating products of scientific investigation, shouldn't his attitude about scientific method be similarly open? Shouldn't the strong claim about preeminence attach only to ideal scientific method, whatever that turns out to be?

Obviously Peirce is not wedded to the details of any present characterization of scientific methodology, but it does seem to me that complete openness to any future characterization of scientific method does render the account vacuous. The details of scientific method may surely be revised in terms of its successes and failures, but if the notion is going to mean something other than "the best method" it must be distinguishable from others by some substantive characterization of a core that perdures over time. If ideal scientific method could look more like Spinoza's than Newton's, what content would be left in the present demarcation of science from other modes of inquiry? Peirce would seem to be committed to his general characterization of the core of scientific method in a way that he is not committed to the content of present scientific theory if he is going to identify our sciences as the legitimate predecessors of the ultimately adequate account.[2]

The issue of scientific realism has been and continues to be a tangled web. The term "realism" itself has a chameleonlike nature, taking on quite different nuances of meaning when the contrast term is nominalism, idealism, or, most recently, anti-realism. Peirce himself always used it in its contrast to nominalism, and his realism issue focused on the status of scientific laws. But the view of science he articulated does have a direct bearing on the contemporary realism debates, even though he did not always frame the relevant questions in terms of his notion of "realism."

On the contemporary scene those who are regarded as scientific realists think in terms of the truth of scientific theories and the reference of its theoretical terms to mind-independent structures, processes, and entities in the world. Non-realists, on the other hand, think in terms of the adequacy of scientific theories and the instrumental usefulness of theoretical constructions in predicting and controlling the phenomena of our empirical world. Peirce's views seem to fit in neither camp. He clearly thinks that the discovery of truth is the goal of science but then defines truth in terms of the ultimate agreement of the scientific community, rather than some correspondence with a mind-independent reality. He clearly thinks of his view as a version of realism, but it is a realism in which the concept of truth is tied to verification and the concept of reality does not invoke the notion of a mind-independent world. Is this just idealism under a different guise, or is it a justifiably realistic picture of the scope and reach of science?

Putnam's views again come to mind, this time his notion of "internal realism." He too invokes a notion of truth that does not involve any correspondence with transcendent structures and a notion of reality or world that is as much made as discovered. His internal realism involves a conception of truth as "idealized rational acceptability"[3] and the thesis that "objects do not exist independent of conceptual schemes."[4] Again, in very Peircean fashion, he views both features of internal realism as following from "the renunciation of the notion of the 'thing-in-itself.'"[5] The point on which I want to focus, however, is a fourth one, namely, Putnam's very general claim that "internal realism is, at bottom, just the insistence that realism is not incompatible with conceptual relativity."[6] It is with regard to this last point that Putnam acknowledges that his internal realism ("I should have called it pragmatic realism"[7]) may well be more like the pragmatisms of James and Dewey than that of Peirce.[8]

Focusing on the claim about conceptual relativity, what exactly is the difference between Peirce's view of scientific explanation and Putnam's internal realism, and what can we say about this difference? In the first place, there are certain kinds of conceptual pluralism that Putnam includes under the rubric 'conceptual relativity', namely, those bearing on different levels of description of the same world, that don't at all divide him from Peirce. Putnam puts the general point this way:

> Realism ... is a view that takes our familiar common-sense scheme, as well as our scientific and artistic and other schemes, at face value without helping itself to the notion of the thing 'in itself'.[9]

Peirce would have no difficulty with this kind of conceptual relativity. In fact, it maps quite neatly onto his own characterization of the relation between our ordinary and our scientific descriptions of the world in terms of different levels of vagueness and precision. Each can be regarded as 'true', but only because they are at different levels; they are not really incompatible. Science is a privileged but not exhaustive description of the world.

Putnam's conceptual relativity, however, cuts deeper than this. He extends it to incompatible theories at the same level of analysis even in the long run. Speaking of theories on the same level of analysis, whether it be physics or psychology, he makes the sweeping claim:

> A twentieth century realist *cannot* ignore the existence of equivalent descriptions: realism is not committed to there being one true theory (and *only* one).... Assuming that there is a 'fact of the matter' as to 'which is true' (if either) *whenever* we have two intuitively 'different' theories, is naive. 'Theories' which differ on which pairs of events are *simultaneous* are certainly 'intuitively different', but after Einstein we know that such 'theories' may, nonetheless, be *equivalent*.... And a sophisticated realist recognizes the existence of equivalent descriptions, because it follows from his theory of the world that there are these various descriptions, as it follows from a geographer's description of the earth that there are alternative mappings (mercator, polar etc.).[10]

What seems to divide Peirce and Putnam here is the idea of scientific convergence and the concept of the ideal limit of scientific investigation that goes along with it for Peirce. The issue is not convergence simply speaking, because Putnam's account includes an element of development and convergence (later scientific accounts being 'better than' earlier ones plus inadequate ones being eliminated) but rather the kind of convergence envisioned. Peirce thinks that the notion of *the* ultimately correct account functions as a regulative ideal vis-à-vis scientific inquiry, thus giving meaning to the notion that there is a way the world is. Putnam entertains a more plastic conception of the world in which there may well be irreducibly plural scientific versions of the world even in the long run. In fact, he implicitly defines his internal realism (to which the contrast term is "metaphysical realism") in terms of the rejection of this regulative ideal: "what makes the metaphysical realist a *metaphysical* realist is his belief that there is somewhere 'one true theory' (two theories which are true and complete descriptions of the world would be mere notational variants of each other)."[11]

According to this definition Peirce's realism is still haunted by the ghost of the metaphysical. Peirce's 'long run', while certainly not the concept of the 'thing-in-itself', plays a role which is a surrogate for that rejected notion, and even this surrogate role Putnam wants to purge. But does he—or can he—succeed, or is there something intractable about this regulative ideal? Putnam himself acknowledges the intractability:

> I can sympathize with the urge behind this view (I would not criticize if I did not feel its attraction). I am not inclined to scoff at the idea of a noumenal ground behind the dualities of experience, even in all attempts to talk about it lead to antinomies. Analytic philosophers have always tried to dismiss the transcendental as nonsense, but it does have an eerie way of reappearing.... Because one cannot talk about the transcendent or even deny its existence without paradox, one's attitude to it must, perhaps, be the concern of religion rather than of rational philosophy.[12]

Putnam views the intractability simply as an invitation to religion, whereas Peirce in a more Kantian fashion also underscores its regulative role in scientific inquiry.

Just as in philosophy of science, there are several ongoing discussions in epistemology to which Peirce's views could make a significant contribution. On the most general level, the kind of particularism his view embodies not only bridges the gap between philosophy of science and epistemology but underscores the public and historical character of knowledge from the very beginning. If the paradigm case of knowing is knowledge 'writ large', namely, the historically established results of science, in one stroke knowledge is located in the public arena, and we are guaranteed an example robust enough for the structure of knowing to be discernible. As a corrective to the Cartesian strategy, this has much to recommend it, at least by way of initial approach to the question of knowledge.

More specifically, his critical common-sensism seems designed to respond to the attractive features of both foundationalism and coherentism on the contemporary scene. There is something that rings true about the view that, given the kind of information processing systems we are, direct perceptual reports should have a privileged status in our belief systems. However, it also seems correct that our grasp of these basic inputs from our environment is not at all infallible and is in fact informed by and conditioned by the other beliefs and expectations we bring to direct experience. The appeals of empiricist foundationalism and holistic coherentism rest on these two insights respectively.

Peirce's overall framework responds to both of these insights simultaneously. On his account there is a range of perceptual beliefs about our world that does have the presumption in its favor and does function decisively in the generation and justification of our overall system of beliefs about the world. However, these empirical inputs, to the degree that they are genuinely epistemic, are shaped to some extent by our concepts and have what justification they have in terms of their roles in our overall epistemic systems. We have no epistemic grasp of 'the simply given', nor is it the case that justification is a matter of mere coherence with our other beliefs. Cognition is a response to an independent empirical world, but it is an active interpretative response that draws on the past and looks to the future. His is a decidedly post-Kantian empiricism that transcends most of the traditional objections to empiricism while retaining the intuition that for us the empirical inputs from our environment are epistemically focal. This kind of epistemological outlook—together with the attention to science as a paradigm case of knowing—is a philosophical orientation with much to recommend it.

This brings me finally to the issues in the philosophy of

mind. While there are philosophers on the contemporary scene who resonate positively to Peirce's very speculative suggestions concerning the semiotic conception of the person, I will focus on his more specific claims about a language of thought and about the irreducibly subjective dimension of human experience. The language-of-thought thesis has had a long and distinguished history and is still very much alive not only in philosophy but in speculative psychology circles today. To the degree that one thinks of mental processes as at all computational, one would seem to be committed to a medium of computation, some kind of internal representational system. Moreover, on the assumption that we don't have direct cognitive access to this internal representational system, its status would be that of a postulated structure necessary to explain a certain order of behavior, and the model for such a structure would be the public representational system with which we are most familiar. Hence, the basic approach to fundamental issue in the philosophy of mind would not be introspection but scientific theory construction. As we have seen, Peirce was one of the most reflective spokesmen for this tradition.

Once one posits a language of thought, however, there seem to be two very different paths one can take in developing the account of thinking (and other mental activity) in which it functions. The first way is that characterized by Putnam as the picture of the mind as a cryptographer:

> The mind thinks its thoughts in mentalese, codes them in the local natural language, and then transmits them (say by speaking them out loud) to the hearer. The hearer has a cryptographer in his head too, of course, who thereupon proceeds to decode the "message." In this picture natural language, far from being essential to thought, is merely the vehicle for the communication of thought.[13]

This Cartesian way of developing the account invites such features as innatism and a private language, and provides a framework for a mentalistic account of both intentionality and justification. Putnam, by invoking a public and holistic account of meaning, goes on to argue against this mentalistic picture of language wherein mental episodes are construed as semantic representations.

But this is not the only way to go with a language-of-thought thesis and, given Peirce's pervasive anti-Cartesianism, we would not expect it to be his way. For Peirce, the language of thought is invoked as part of an account of the paradigmatic public 'thoughts' of the scientific community. Public language is not merely a medium for the communication of thought but, as embedded in other concrete practices, the paradigmatic instance of thought itself. Public communication does not merely *express* thought but *is* thought, and the 'inner episodes' do not function primarily semantically or epistemically, since the beliefs have both their meaning and their justification in virtue of public factors. The 'inner episodes' are theoretical posits that play a role in explanation but not in analysis or justification. This version of the language of thought thesis responds to the pull of both the externalist and the internalist accounts of our mental lives and provides a framework for a truly non-reductive philosophy of mind.[14]

This non-reductive orientation carries over to his account of the subjective dimension of human experience. There certainly is an external and objectivist thrust to Peirce's attempt to understand our mental life in terms of a system of representations that progressively approximate to the ideal representation of the world. This account is broadly speaking functionalist, inasmuch as it is in terms of a system of signs that can be instantiated in a wider and wider community of rational beings that need not even be human. Truth is conceived of in terms of the convergence of that community on a unique representation of the world as the idosyncracies of individual perspective are washed out over time.

For Peirce, however, this objectivist account of the mental is supplemented by an account of consciousness as that irreducible subjectivity that he finds in human experience and that he identifies with an organically based range of feeling. These two dimensions being acknowledged and being acknowledged to be ineliminable, the complete objective account of the world should not only involve a system of objective representations but also place-holders for the plurality of finite perspectives on this world.

Peirce does not satisfactorily work out the details of this synoptic vision of the world, but he can hardly be blamed for that. Dissatisfied with the details of the heroic solution proffered by his great predecessor, Kant, he showed that he grasped the full scope of the problem and pointed us in the direction of a socio-historical conception of science that would seem to be crucial to its satisfactory resolution. This may well be the best one can do.[15] Fashioning a conception of the world with our place in it is *the* human speculative project, and we have overwhelming inductive evidence to the effect that it is a project that has to be attempted anew for every age. The correct account of the world with our place in it at best functions as the ideal limit of our feeble efforts, although it is perennial sign of our hubris that we each tend to think of it as a description of our present offering.

Notes

I. Introduction: Life and Work

1. Christopher Hookway, *Peirce* (London: Routledge & Kegan Paul, 1985), p. 1.

2. While we await Max Fisch's anticipated biographical study of Peirce, information about his life can be culled from various sources, among them the following: Max Fisch, "Introduction," in *Writings of Charles S. Peirce: A Chronological Edition*, 4 vols. (Bloomington: Indiana University Press, 1982–89). Max Fisch, "Was There a Metaphysical Club in Cambridge?" in *Studies in the Philosophy of Charles Sanders Peirce*, ed. Moore and Robin (Amherst: University of Massachusetts Press, 1964). Carolyn Eisele, "Introduction," *The New Elements of Mathematics by C. S. Peirce*, 4 vols. (Atlantic Heights, N.J.: Humanities Press, 1976). Carolyn Eisele, "Peirce the Scientist," in *Historical Perspectives on Peirce's Logic of Science: A History of Science*, ed. Carolyn Eisele (Amsterdam: Mouton, 1985). Carolyn Eisele, "A Short Scientific Biography" in *Studies in the Scientific and Mathematical Philosophy of Charles S. Peirce*, by Carolyn Eisele (The Hague: Mouton, 1979). Victor Lenzen, "Charles S. Peirce as Astronomer" in *Studies in the Philosophy of Charles Sanders Peirce*, ed. Moore and Robin (Amherst: University of Massachusetts Press, 1964). Victor Lenzen, "Charles S. Peirce as Mathematical Geodesist" in *Transactions of the Charles S. Peirce Society* 8, no. 2 (1972): 90–105. Victor Lenzen, "Charles S. Peirce as Mathematical Physicist," *Transactions of the Charles S. Peirce Society* 10, no. 3 (1975): 159–66.

II. Peirce's Account of Science

1. This conceptualization of the three forms of reasoning as three stages of scientific inquiry is the mature development of Peirce's earlier explications of the three forms of inference in terms of the three figures of the syllogism and in terms of three different patterns involving Rules, Cases and Results. The mature view did not receive

its fully explicit formulation until 1901 (7.202–7), but Peirce saw it as continuous with the earlier expressions. On this point see K. T. Fann, *Peirce's Theory of Abduction* (The Hague: Martinus Nijhoff, 1970).

2. Cf. 7.139–61. For a more sustained treatment of this feature of Peirce's philosophy of science see chapter 4 of Nicholas Rescher, *Peirce's Philosophy of Science* (Notre Dame, Ind.: University of Notre Dame Press, 1978).

3. It is important to note that Peirce is here rejecting only the general method of authority for fixing belief. He is not claiming that there is no role for authority in what he characterizes as the scientific method of fixing belief. In fact, authority properly construed plays an important role in his account of scientific method. For his positive attitude toward authority see section 5 of this chapter.

4. This is the construal offered by Thomas Knight, *Charles Peirce* (New York: Washington Square Press, 1965), p. 30.

5. For an excellent discussion of this dimension of abduction see Timothy Shanahan, "The First Moment of Scientific Inquiry: C. S. Peirce on the Logic of Abduction," in *Transactions of the Charles S. Peirce Society* 22 (Fall 1986): 449–66.

6. Peirce is simply articulating a necessary condition for the general grounding of scientific inquiry. He is not committed to the view that all dimensions of scientific speculation, even the more abstract, are specifically grounded in such a survival instinct. In fact, he explicitly disavows this broader claim for instinct: "When we theorized about molar dynamics we were guided by our instincts. Those instincts had some tendency to be true because they had been formed under the influence of the very laws they were investigating. But as we penetrate further and further from the surface of nature, instinct ceases to give any decided answer" (7.508). But our instincts get us started in the right direction. At the more abstract reaches of science we have to make do with the other constraints on theorizing.

7. Nicholas Rescher develops an interesting economic critique of Peirce's scientific realism to the effect that, given the way science develops and our economic limits, there will be a dimension of the theoretically accessible world that will be de facto inaccessible. This, however, does not count against the definitional point. Cf. N. Rescher, *Peirce's Philosophy of Science* (Notre Dame, Ind.: University of Notre Dame Press, 1978), pp. 33–39.

8. For an interesting discussion of Peirce's views on the reality of law see Bas van Fraassen, *Laws and Symmetry* (Oxford: Clarendon Press, 1989), pp. 19–23.

9. The exegetical question is far more complicated than my use of "he saw" conveys. At one point, while acknowledging that "perhaps the writer wavered a bit in his own mind," he goes on to point out that the text of the original treatment was replete with the "would be's" characteristic of the later account (5.453), and in another place he explicitly retracts his own characterization of his earlier position as nominalistic (5.403, n. 3). The text itself seems to support both his own self-criticism and the retraction thereof.

10. Peirce was one of the earliest philosophers to worry about the status of counterfactuals, and his reflections on them are much more subtle and complicated than these brief paragraphs convey. For a more extensive treatment see Peter Skagestad, *The Road of Inquiry* (New York: Columbia University Press, 1981), pp. 98–117.

11. The view of perception and perceptual belief involved here will be developed in greater detail in chapters 3 and 4.

12. Arthur Smullyan, "Some Implications of Critical Common-Sensism," in *Studies in the Philosophy of Charles Sanders Peirce*, ed. Wiener and Young (Cambridge, Mass.: Harvard University Press, 1952), p. 118.

13. To this comment about reasoning he immediately adds: "Now if exactitude, certitude and universality are not to be attained by reasoning, there is certainly no other means by which they can be reached." This more general epistemological claim will be discussed later.

14. This was included in the *Report of the Superintendent of the U.S. Coast Survey* (1873) (*W*, 3.114–60).

15. Peirce does consider the perennial reflexive challenge—is the statement of fallibilism itself fallible or is it infallible? He clearly considers this a cute point not to be taken all that seriously but does add that if you want to insist on it he will reformulate his claim to the effect that every statement except the statement of fallibilism is fallible: "If I must make an exception, let it be that the assertion that 'every assertion but this is fallible', is the only one that is absolutely infallible" (2.75).

III. Peirce's Critique of Cartesian Epistemology

1. Peirce's use of the term "faculty" here is not casual. He has a reasoned defence of the employment of the notion of faculties in psychological analysis: "If there is to be any positive science of psychology—that is the science of the mind itself—it must be that the mind is more than a *unit*. It must have parts, and these will be

faculties. Each of these faculties will have special functions and these functions will be simple conceptions. We can only know faculties, however, through their functions; accordingly, the knowledge of simple conceptions will be the knowledge of the mind itself and the analysis of conceptions will be psychology" (*W*, 1.63–64). Whether or not Peirce's employment of this methodological strategy avoids the notorious difficulties of 'faculty psychology' depends in large measure on his account of the sense in which the mind can be said to have 'parts' and the kind of explanation this way of conceptualizing things purports to make available.

2. The text for this argument is 5.263, but it is informative to look at other slightly different formulations in "Questions on Reality" (*W*, 2.178) and in "Potentia ex Impotentia" (*W*, 2.191).

3. The class of indubitable beliefs for Peirce is much broader than the perceptual judgments on which I am focusing, but I am attending merely to the latter because my principal interest here is in the structure of empirical knowledge. In addition to indubitable perceptual judgments, Peirce also calls attention to some indubitable beliefs of a very general and recurrent kind, as well as indubitable principles of inference. Cf. 5.441–42.

4. Otto Neurath, "Protocol Sentences" in A. J. Ayer, ed., *Logical Positivism* (New York: The Free Press, 1959), p. 201.

5. Karl Popper, *The Logic of Scientific Discovery* (New York: Basic Books, 1959), p. 111. In fact, Popper's metaphor is exactly the one used by Peirce. Speaking of the role of *il lume naturale* in the progress of science, Peirce invokes the metaphor: "The only end of science as such is to learn the lesson that the universe has to teach it. In Induction it simply surrenders itself to the force of facts. But it finds at once–I am partly inverting the historical order, in order to state the progress in its logical order—it finds I say that this is not enough. It is driven in desperation to call upon its inward sympathy with nature, its instinct for aid, just as we find Galileo at the dawn of modern science making his appeal to *il lume naturale*. But insofar as it does this, the solid ground of fact fails it. It feels from that moment that its position is only provisional. It must then find confirmations or else shift its footing. Even if it does find confirmations, they are only partial. It is still not standing on the bedrock of fact. It is walking upon a bog, and can only say, this ground seems to hold for the present. Here I will stay until it begins to give way. Moreover, in all its progress, science vaguely feels that it is only learning a lesson" (5.589).

6. Willard V. O. Quine, "Two Dogmas of Empiricism" in *From a Logical Point of View* (New York: Harper Torchbook, 1963), pp. 42–43.

7. Strictly speaking, the matter is a little more complicated than my interpretation suggests. What I am calling 'perceptual experience' does not take account of Peirce's distinction between the concept of experience and the concept of perception: "It is more particularly to changes and contrasts of perception that we apply the word "experience." We experience vicissitudes, especially. We cannot experience the vicissitude without experiencing the perception which undergoes the change; but the concept of *experience* is broader than that of *perception*, and includes much that is not, strictly speaking, an object of perception. It is the compulsion, the absolute constraint upon us to think otherwise than we have been thinking that constitutes experience" (1.336). However, given the specific set of issues I am exploring, the elision embodied in the notion of perceptual experience simplifies without falsifying the issue at hand.

8. For an excellent discussion of this and related points see Christopher Hookway, *Peirce* (London: Routledge & Kegan Paul, 1985), pp. 127–41; 166-72.

IV. Mind and Reality

1. Since I am not here interested in his theory of signs or his logic per se but only insofar as they inform his philosophy of mind, the full generality of his theories of signs and of inference will not be explored, but only those dimensions of them that bear on his philosophy of mind. For a good discussion of his theory of signs see Douglas Greenlee, *Peirce's Concept of Sign* (The Hague: Monton Publishers, 1973).

2. Characteristically, Peirce distinguished three kinds of consciousness corresponding to his three categories, i.e., immediate feeling, the feeling of reaction, and the feeling of habit formation or learning (1.382–85). However, this level of distinction does not bear on the general point being discussed here. For a good discussion of this see Nathan Houser, "Peirce's General Taxonomy of Consciousness," *Transactions of the Charles S. Peirce Society* 19 (Fall 1983): 331–59.

3. Peirce does little more than locate the problem of self-control: "The power of self-control is certainly not a power over what one is doing at the very instant the operation of self-control is commenced.

It consists (to mention only the leading constituents) first, in comparing one's past deeds with standards, second, in rational deliberation concerning how one will act in the future, in itself a highly complicated operation, third, in the formation of a resolve, fourth, in the creation, on the basis of the resolve, of a strong determination or modification of habit. This operation of self-control is a process in which logical sequence is converted into mechanical sequence or something of the sort. How this happens, we are in my opinion as yet entirely ignorant" (8.320).

4. Houser, "Peirce's General Taxonomy of Consciousness," pp. 341–43.

5. For a summary of some of his many arguments against determinism see 6.64. For a critical reaction to Peirce's identification of materialism and mechanism see Larry Holmes, "Peirce's Philosophy of Mind," in *Studies in the Philosophy of Charles Sanders Peirce*, ed. Edward Moore and Richard Robin (Amherst: University of Massachusetts Press, 1964), p. 368.

V. Conclusion

1. Hilary Putnam, *The Many Faces of Realism* (LaSalle, Ill.: Open Court, 1987), p. 72.

2. An alternative substantive characterization of scientific method might be to think not in terms of a perduring core, however general, but of a series of changing methods unified by an overlapping family resemblance, which would constitute a unique chain through time. It is not clear that this conception would avoid the difficulty because, as the color spectrum illustrates, the ideal scientific method could be completely different from our present one, more like some other method than present science, and present science might not even be the only way to get there from here. Hence, our present demarcation of science from other modes of inquiry would still be vacuous.

3. Hilary Putnam, *Representation and Reality* (Cambridge: MIT Press, 1988), p. 115 and Hilary Putnam, *Reason, Truth and History* (Cambridge: Cambridge University Press, 1981), p. 55.

4. Putnam, *Reason, Truth and History*, p. 52.

5. Putnam, *The Many Faces of Realism*, p. 36.

6. Ibid., p. 17.

7. Ibid.

8. Cf. Ibid., p. 70.

9. Ibid., p. 17.

10. Hilary Putnam, *Meaning and the Moral Sciences* (London: Routledge & Kegan Paul, 1978), pp. 50–51.

11. Hilary Putnam, *Realism and Reason* (Cambridge: Cambridge University Press, 1983), p. 211.

12. Ibid., p. 226.

13. Hilary Putnam, *Representation and Reality*, pp. 6–7.

14. The recent philosopher who has developed most fully the details of an account along these lines is Wilfrid Sellars. See Wilfrid Sellars, *Science, Perception and Reality* (London: Routledge & Kegan Paul, 1963), pp. 177–96; and C. F. Delaney et al., *The Synoptic Vision: Essays on the Philosophy of Wilfrid Sellars* (Notre Dame, Ind.: University of Notre Dame Press, 1977), pp. 15–31.

15. The most recent attempt to think through these questions at the deepest level is Thomas Nagel's *The View from Nowhere* (Oxford: Oxford University Press, 1986).

About the Author

C. F. DELANEY, who received his Ph.D. from St. Louis University, is Professor of Philosophy at the University of Notre Dame. He was Chairman of the Philosophy Department at Notre Dame from 1972–82 and is currently Director of the Honors Program. He has been active in many professional organizations, serving as President of the American Catholic Philosophical Association and President of the Charles S. Peirce Society. He has published a number of books, including *Rationality and Religious Belief* (Notre Dame Press, 1979), *The Synoptic Vision* (Notre Dame Press, 1977), and was coeditor of *Science and Reality* (Notre Dame Press, 1984). He is a contributing editor to the *Writings of Charles S. Peirce: A Chronological Edition.*

Index